BIOMEDICAL & NANOMEDICAL TECHNOLOGIES
CONCISE MONOGRAPH SERIES

Biopolymers Based Micro- and Nano-Materials

Nitar Nwe

Co-published by Momentum Press, LLC, 222 E. 46th Street, #203, New York, NY 10017, USA (www.momentumpress.net)

A catalog record is available from the Library of Congress.

Print ISBN: 978-0-7918-6040-3
ASME Order No. 860403
Electronic ISBN: 978-1-60650-659-2

Series Editors' Preface

Biomedical and Nanomedical Technologies (B&NT)
This **concise** monograph series focuses on the implementation of various engineering principles in the conception, design, development, analysis and operation of biomedical, biotechnological and nanotechnology systems and applications. The primary objective of the series is to compile the latest research topics in biomedical and nanomedical technologies, specifically devices and materials.

Each volume comprises a collection of invited manuscripts, written in an accessible manner and of a concise and manageable length. These timely collections will provide an invaluable resource for initial enquiries about technologies, encapsulating the latest developments and applications with reference sources for further detailed information. The content and format have been specifically designed to stimulate further advances and applications of these technologies by reaching out to the non-specialist across a broad audience.

Contributions to *Biomedical and Nanomedical Technologies* will inspire interest in further research and development using these technologies and encourage other potential applications. This will foster the advancement of biomedical and nanomedical applications, ultimately improving healthcare delivery.

Table of Contents

Abstract

Nowadays biopolymers such as agar, agarose, alginate, carrageenan, cellulose, chitin, chitosan, collagen, hyaluronic acid, gelatin, glucan, starch, DNA, RNA and protein have been produced from laboratory to industrial scale. The physico-chemical properties of these biopolymers such as chemical compositions, solubility, molecular weight and viscosity are of major importance on the preparation of micro- and nanomaterials such as powder, solution, hydrogel, micro- and nano-sized pores in membrane, micro- and nano-fiber, macro- and micro-beads, nano-particles, and micro- and nano-structured scaffold. These micro- and nano-materials have been used in various sectors such as agriculture, food, medicine, etc. This monograph will address the source and production methods of biopolymers, properties of biopolymers, preparation of micro- and nano-materials using biopolymers, characterization of micro- and nano-biomaterials and application of micro- and nano-biomaterials.

1. Introduction

Polymers are macromolecules composed with one or more types of repeating monomer units. Biopolymers are polymers synthesized during metabolic process of living organisms. Living organisms are mainly divided into two categories: microorganisms and macroorganisms. Macroorganisms are multicellular organisms (i.e., animal, plants, some fungi and some algae), but microorganisms are unicellular (bacteria, some yeast, some fungi, some algae, etc.) or multicellular (some fungi, some yeasts, some algae, etc) organisms. There are two basic types of cells: prokaryote and eukaryote. Prokaryotic cells are usually unicellular organisms, while eukaryotic cells are composed either unicellular or multicellular organisms. Prokaryotic and eukaryotic cells are mainly synthesized carbohydrates, DNA, RNA, proteins and lipids. Nowadays biopolymers such as polysaccharide, DNA, RNA, proteins and lipids have been produced from macro- and micro-organisms.

Polysaccharide such as agar, alginate and carrageenan are produced from algae. Agarose is produced from agar. Nanofibrillated cellulose is mainly produced from wood pulp and cotton. Moreover microbial cellulose has also been produced by various species of bacteria. Chitin and chitosan are found as supporting materials in many aquatic organisms, in many insects, in terrestrial crustaceans, in mushrooms, in some fungi and in some yeast. Commercially chitin and chitosan are produced form aquatic organisms. Commercially starch is produced from cassava root, sweet potato, rice, potato, wheat and corn. Plasmid DNAs is produced by fermentation process followed by chromatographic purification method. Proteins such as collagen and gelatin have been produced from ocean fish's skin, pork skin and cow bone.

Biopolymers have been used to prepare various forms of micro- and nano-materials and these materials have been applied in many food products, medical products, etc. This monograph describes the production and characterization of biopolymers, preparation of macro-, micro- and nano-materials using biopolymers and their applications.

2. Production of biopolymers

2-A Production of agar

Agar has been extracted in a variety of ways including dissolve agar from seaweeds in hot water and extract agar from agar solution (Figures 2-1 and 2-2). Most researchers believed that agar is composed with agarose and agaropectin.

2-B Production of agarose

Agarose is manufactured from agar (Figure 2-3).

2-C Production of alginate

Commercially alginates are produced by extraction from brown seaweeds such as *Laminaria digitata, Laminaria hyperborea* and *Macrocystis pyrifera* and from several bacteria such as the nitrogenfixing aerobe *Azotobacter vinelandii* and the opportunistic pathogen *Pseudomonas aeruginosa* also produce alginate (Gorin and Spencer 1966, and Evans and Linker 1973 cited in [4]). Production process of alginate from *Azotobacter vinelandii* is shown in Figure 2-4. Moreover production of alginate from seaweed in industrial scale is shown in Figure 2-5.

2-D Production of carrageenan

There are two basic processes (Alcohol precipitation and KCl precipitation) used for the production of carrageenan. The traditional process for the production of carrageenan is the alcohol precipitation process (Figure 2-6).

2-E Production of cellulose: Plant and bacterial source

Bacterial cellulose has been produced from bacteria, *Acetobacter xylium* in medium containing glucose, 8 g; yeast extract, 2 g and distilled water 200 ml at 30°C for 7 days (Figure 2-7).

Fermentation medium inoculated with subculture *A. xylum* placed in a rubber mold and incubated at 30°C for 7 days. It was observed that bacterial cellulose formed entire shape of the rubber mold (Figure 2-8). However bacterial cellulose formed upper surface of plastic or glass mold when *A. xylum* was incubated in medium placed in plastic or glass mold (Figure 2-7). Bacterial cellulose can be produced in various forms by cultivation of *A. xylum* in different shape of molds (Figure 2-8).

Cellulose gel obtained from bacteria is not pure. It contains some impurities such as bacteria, medium components, etc. The impurities in bacterial cellulose were removed by treatment with NaOH or KOH or Na_2CO_3 at

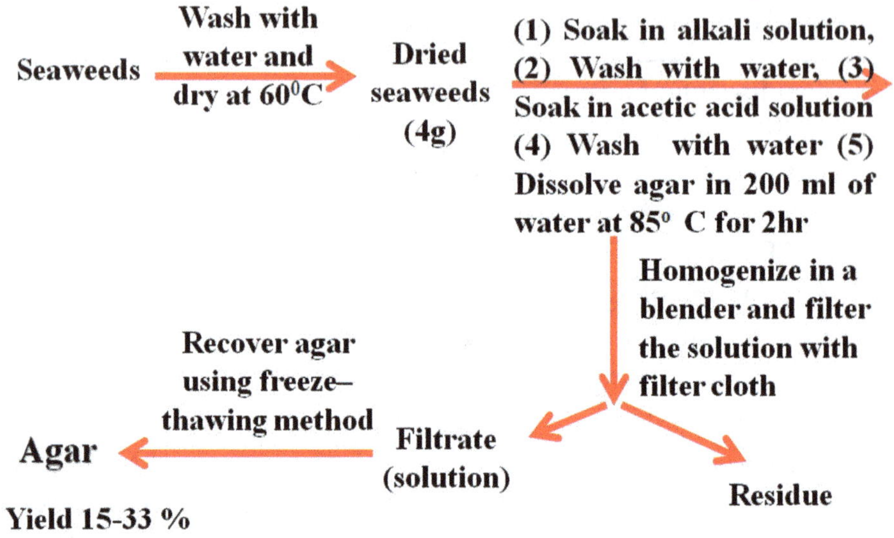
Seaweeds
Wash with water and dry at 60⁰C
Dried seaweeds (4g)
(1) Soak in alkali solution, (2) Wash with water, (3) Soak in acetic acid solution (4) Wash with water (5) Dissolve agar in 200 ml of water at 85° C for 2hr
Homogenize in a blender and filter the solution with filter cloth
Filtrate (solution)
Residue
Recover agar using freeze–thawing method
Agar
Yield 15-33 %

Figure 2-1 Flow chart for production of agar in laboratory scale [1].

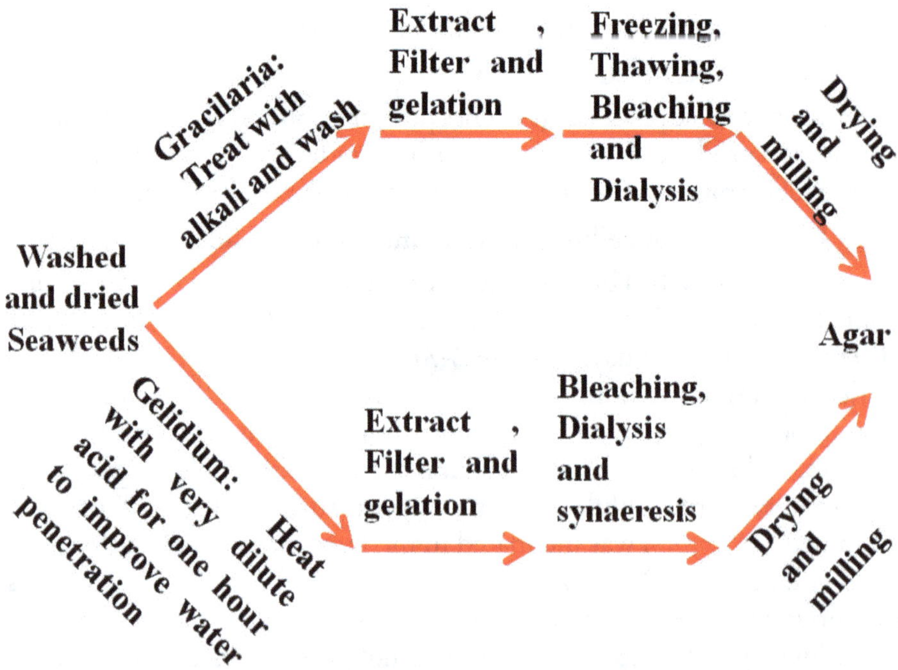
Washed and dried Seaweeds
Gracilaria: Treat with alkali and wash
Extract , Filter and gelation
Freezing, Thawing, Bleaching and Dialysis
Drying and milling
Agar
Gelidium: Heat with very dilute acid for one hour to improve water penetration
Extract , Filter and gelation
Bleaching, Dialysis and synaeresis
Drying and milling

Figure 2-2 Flow chart for production of agar in industrial scale [2].

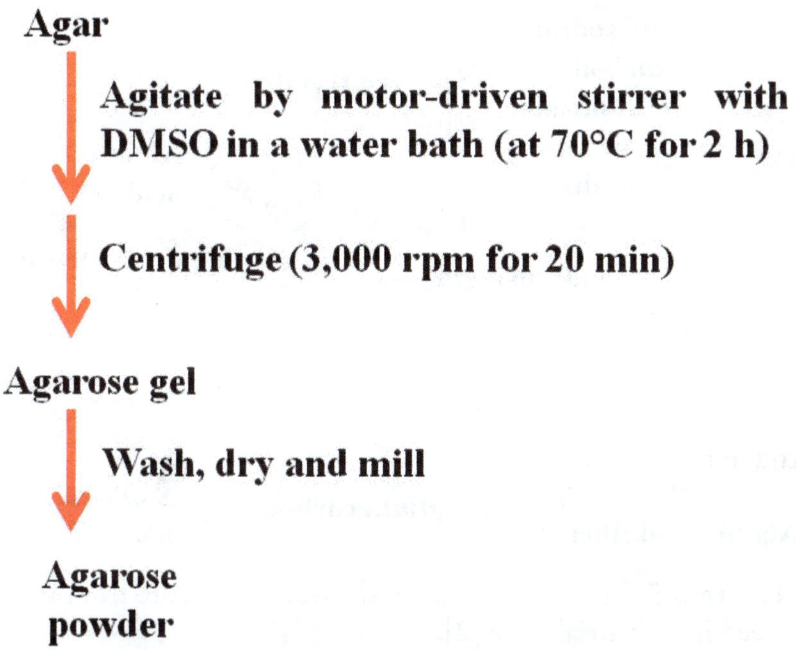
Agar
Agitate by motor-driven stirrer with DMSO in a water bath (at 70°C for 2 h)
Centrifuge (3,000 rpm for 20 min)
Agarose gel
Wash, dry and mill
Agarose powder

Figure 2-3 Flow chart for production of agarose [3].

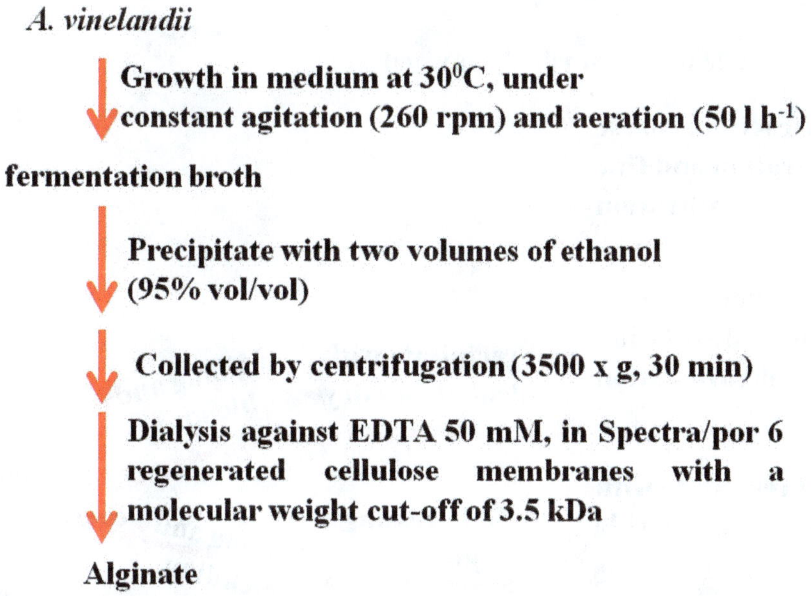
A. vinelandii
Growth in medium at 30^{0}C, under constant agitation (260 rpm) and aeration (50 l h^{-1})
fermentation broth
Precipitate with two volumes of ethanol (95% vol/vol)
Collected by centrifugation (3500 x g, 30 min)
Dialysis against EDTA 50 mM, in Spectra/por 6 regenerated cellulose membranes with a molecular weight cut-off of 3.5 kDa
Alginate

Figure 2-4 Flow chart for production of alginate from bacteria [4].

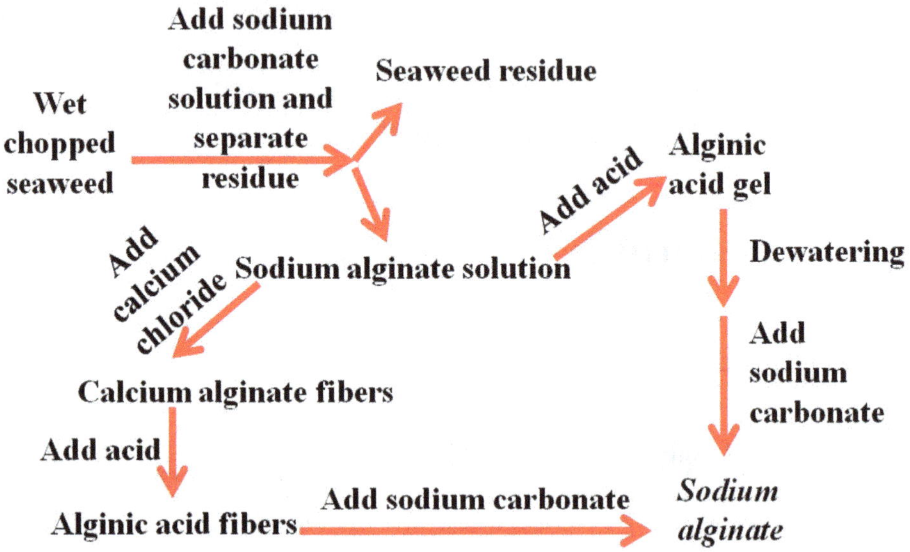
Add sodium carbonate solution and separate residue
Seaweed residue
Wet chopped seaweed
Alginic acid gel
Add acid
Add calcium chloride
Sodium alginate solution
Dewatering
Add sodium carbonate
Calcium alginate fibers
Add acid
Add sodium carbonate
Sodium alginate
Alginic acid fibers

Figure 2-5 Flow chart for production of alginate from seaweed in industrial scale [2].

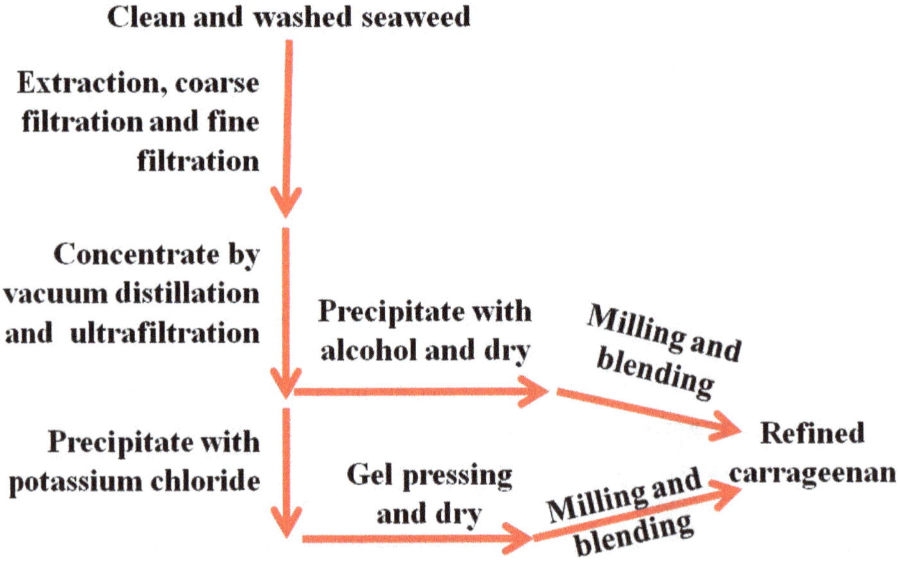
Clean and washed seaweed
Extraction, coarse filtration and fine filtration
Concentrate by vacuum distillation and ultrafiltration
Precipitate with alcohol and dry
Milling and blending
Refined carrageenan
Precipitate with potassium chloride
Gel pressing and dry
Milling and blending

Figure 2-6 Flow chart for production of carrageenan from seaweed in industrial scale [2].

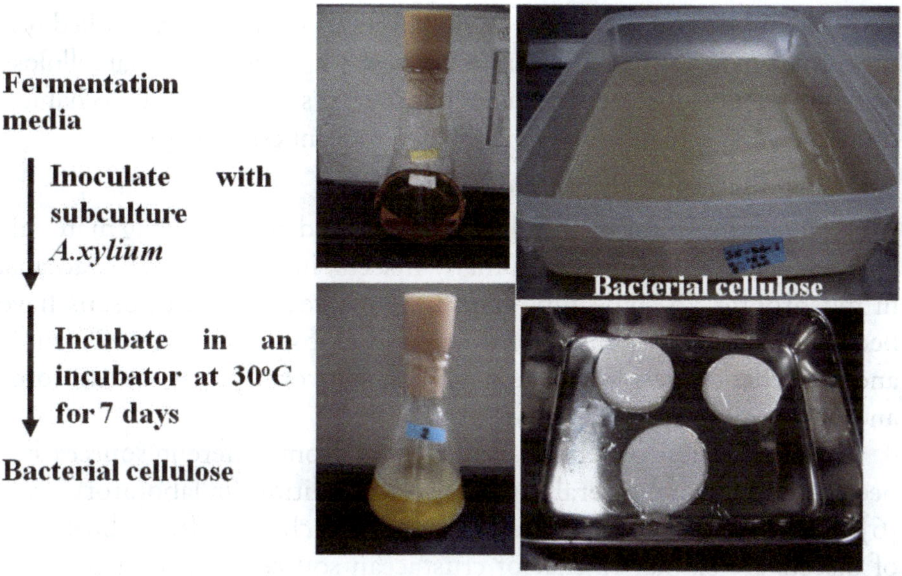
Fermentation media
Inoculate with subculture *A.xylium*
Incubate in an incubator at 30°C for 7 days
Bacterial cellulose
Bacterial cellulose

Figure 2-7 Production of bacterial cellulose from *Acetobacter xylium*.

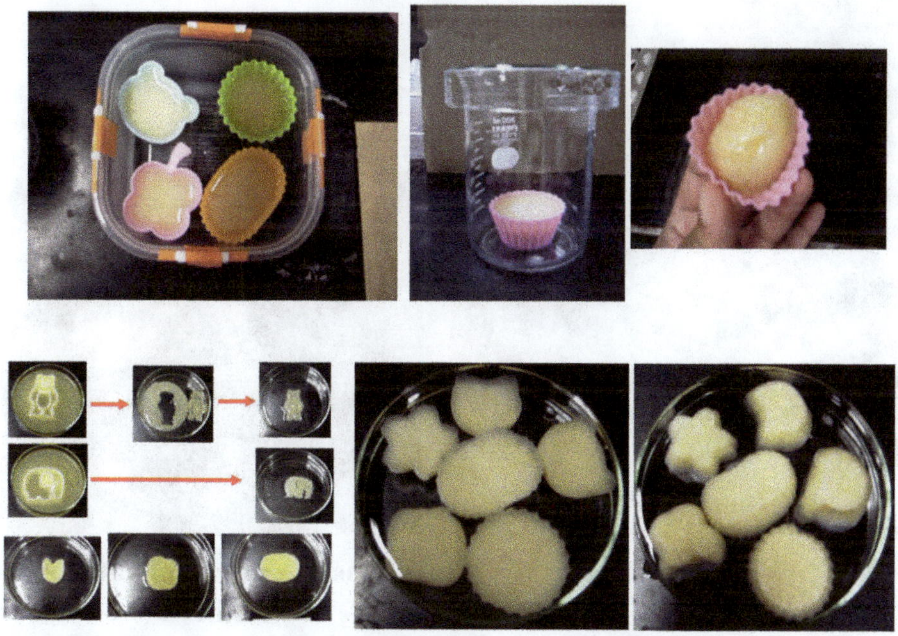

Figure 2-8 Production of various shapes of bacterial cellulose using *Acetobacter xylium*.

100°C for 15–20 min and then cellulose gel was washed with distilled water upto neutral pH [5]. The market price of 1 kg of wet bacterial cellulose was 1000 yen [5]. Bacterial cellulose has unique structural and mechanical properties, and is highly pure as compared to plant cellulose [5].

2-F Production of chitin and chitosan

Natural polymers, chitin and chitosan are found as supporting materials in many aquatic organisms, in many insects, in terrestrial crustaceans, in mushrooms, in some fungi and in some yeast [6]. Chitosans have been isolated from shells of shrimps and crabs, bone plates of squids and cuttlefishes, dried dead honey bees, silkworm pupae, mushrooms and fungal mycelia in laboratory-scale.

The production of chitin and chitosan from different sources has been reported using several experimental conditions in laboratory-scale [6]. The processes for extraction of chitin and chitosan from the cuticle of insects are similar to that of crustacean sources. However pigment compounds in cuticle of insect could not be removed easily, but pigment compound in the shells of crab and shrimp could be removed easily using 1 M NaOH solution (Figure 2-9).

Crab Chitin **Insect cuticle in NaOH solution**

Figure 2-9 Crab chitin obtained from Koyo Chemical Co., Ltd, Japan and insect cuticle treated with 1 M NaOH solution.

Moreover the processes for extraction of chitosan from the cell wall of mushroom are similar to that of fungal mycelia [6]. It is because of organic matrix of aquatic crustacean and insect is different from cell walls of mushroom and fungus (Figures 2-10, 2-11 and 2-12).

According to published data and our research data, chitosan and glucan are the main component of the cell walls of fungi (Figure 2-12). In

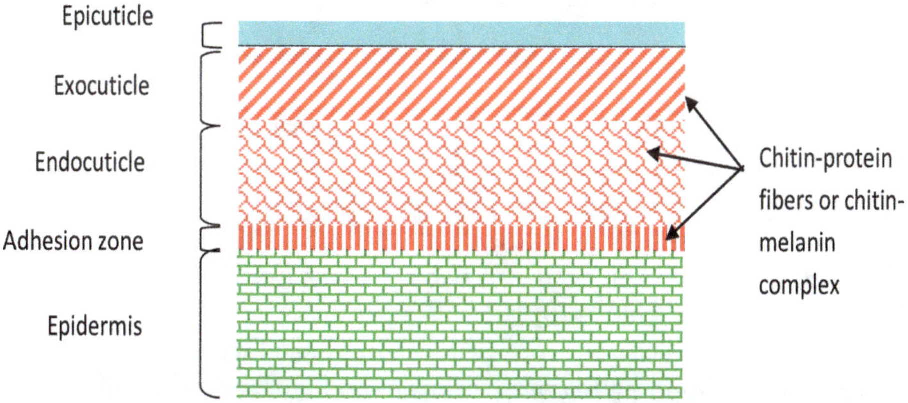
Epicuticle
Exocuticle
Endocuticle
Adhesion zone
Epidermis
Chitin-protein fibers or chitin-melanin complex

Figure 2-10 Schematic interpretation of organic matrix of insect cuticle (not drawn to scale) (Reproduced from Nwe et al., 2011, Chitin, Chitosan, Oligosaccharides and Their Derivatives: Biological Activities and Applications (pp. 3–10). CRC press, Taylor & Francis Group, New York. Reproduced with permission.) [7].

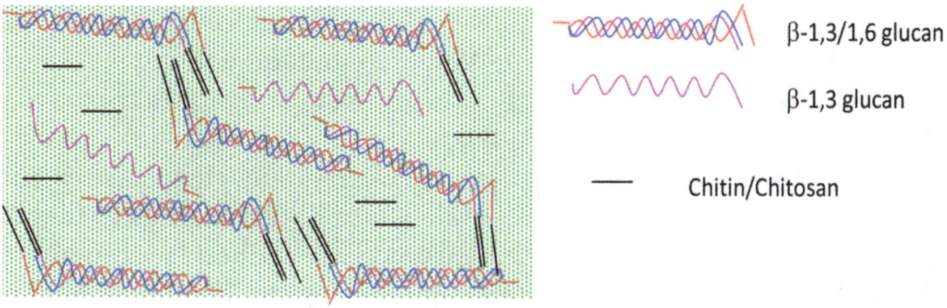
β-1,3/1,6 glucan
β-1,3 glucan
Chitin/Chitosan

Figure 2-11 Structure of cell wall of a typical mushroom (not drawn to scale) (Reproduced from Nwe et al., 2011, Chitin, Chitosan, Oligosaccharides and Their Derivatives: Biological Activities and Applications (pp. 3–10). CRC press, Taylor & Francis Group, New York. Reproduced with permission.) [7].

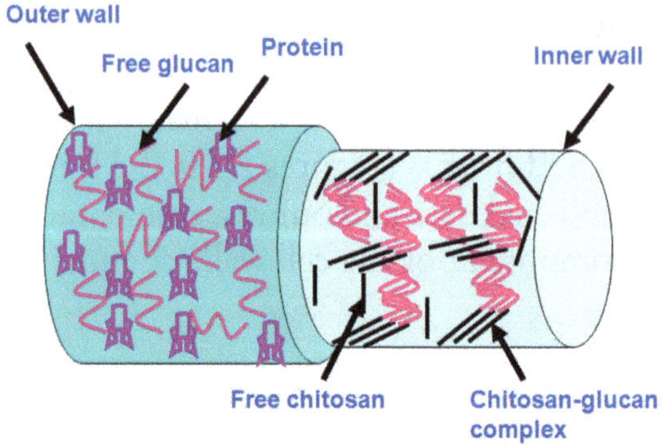
Outer wall
Free glucan
Protein
Inner wall
Free chitosan
Chitosan-glucan complex

Figure 2-12 Model of inner and outer wall of fungal hypha (Reproduced from Nwe et al., 2011, Adv Polym Sci. 244: 187–208. Reproduced with permission.) [8].

their cell walls, chitosan occurs in two forms, as free chitosan and covalently bonded to β-glucan [9, 10].

Within last 14–15 years, our research has been focused on (1) investigation of the bonding between chitosan and glucan in the cell wall of fungus *G. butleri*, (2) comparison the yield of chitosan from mycelia of fungi, *G. butleri* and *A. coerulea*, and (3) applications of fungal chitosan in agriculture and medical sectors [9]. In our research, we found out that chitosan in the fungal cell wall exists in two forms, free chitosan and chitosan bounded to glucan [9, 10]. The linkage between chitosan and β-glucan in the chitosan–glucan complex has been successfully cleaved using a heat stable α-amylase and the resultant chitosan and β-glucan polymeric moieties have been purified and characterized. Data from elementary analysis, IR and ^{13}C NMR spectroscopy, and the results of various enzymatic treatments and reducing sugar analysis confirmed that the β-linked chitosan and β-linked glucan are linked with α-(1-4) glycosidic bond [8, 9, 10]. Based on this observation, enzymatic chitosan extraction method has been developed to obtain high yield fungal chitosan in very easy way [9, 10]. This method has been proposed for the large scale production of fungal chitosan (Figure 2-13). An effective chitosan extraction procedure is essential for an economical production of fungal chitosan [8, 9, 10].

In agriculture and medical applications, low-molecular-weight (LMW) chitosan and chito-oligosaccharide (CTS-O) are more effective than high-molecular-weight (HMW) chitosan [6]. The LMW chitosan and

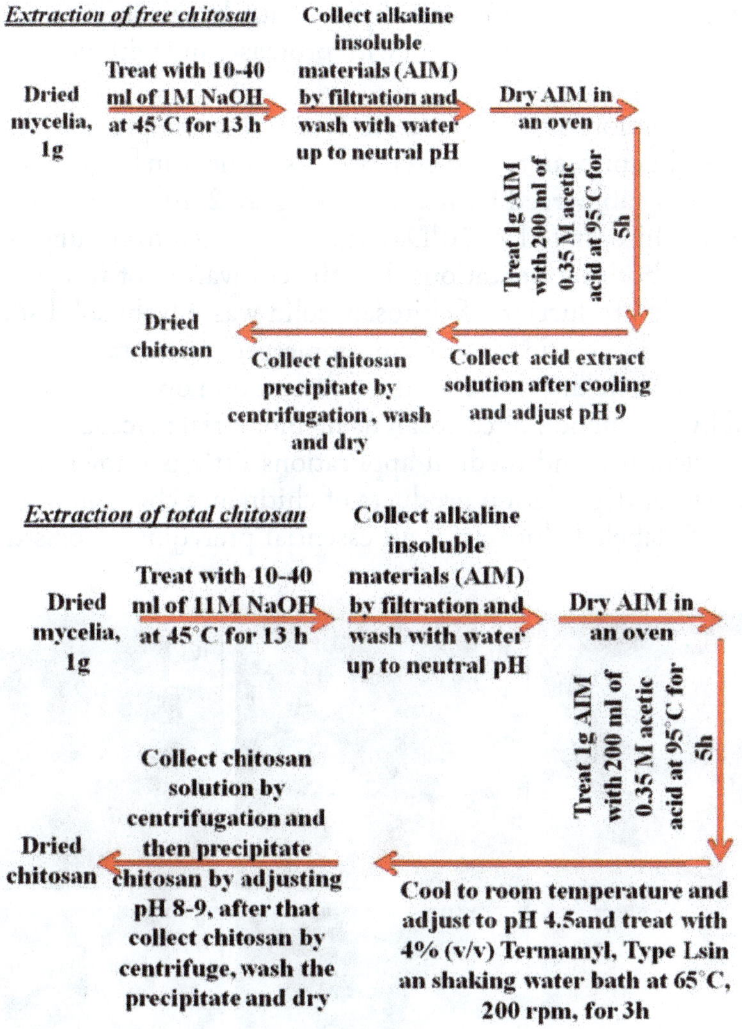
Extraction of free chitosan
Dried mycelia, 1g
Treat with 10-40 ml of 1M NaOH at 45°C for 13 h
Collect alkaline insoluble materials (AIM) by filtration and wash with water up to neutral pH
Dry AIM in an oven
Treat 1g AIM with 200 ml of 0.35 M acetic acid at 95°C for 5h
Collect acid extract solution after cooling and adjust pH 9
Collect chitosan precipitate by centrifugation, wash and dry
Dried chitosan
Extraction of total chitosan
Dried mycelia, 1g
Treat with 10-40 ml of 11M NaOH at 45°C for 13 h
Collect alkaline insoluble materials (AIM) by filtration and wash with water up to neutral pH
Dry AIM in an oven
Treat 1g AIM with 200 ml of 0.35 M acetic acid at 95°C for 5h
Cool to room temperature and adjust to pH 4.5and treat with 4% (v/v) Termamyl, Type Lsin an shaking water bath at 65°C, 200 rpm, for 3h
Collect chitosan solution by centrifugation and then precipitate chitosan by adjusting pH 8-9, after that collect chitosan by centrifuge, wash the precipitate and dry
Dried chitosan

Figure 2-13 Production of free chitosan and total chitosan from fungal mycelia [9, 10].

CTS-O have been produced from the HMW chitosan by physical method such as treatment with γ-ray irradiation and with microwave irradiation; chemical method such as treatment with dilute and concentrated HCl, with $NaNO_2$, with H_2O_2, with HNO_2, and with phosphoric acid; mechanical method such as treatment with sonication; and enzymatic method such as treatment with cellulose, with lysozyme, with chitinase, with chitosanase, with lipase, with hyaluronidase, with papain, with pectinase, with pepsin, with protease, and with hemicellulase [6, 11].

The production of chitosan from fungal sources provides low molecular weight chitosan in two-step process. Fungi can be grown in solid substrate and submerged fermentation (Figure 2-14). Here low molecular weight chitosan (MW 10^4Da) can be obtained from fungus mycelia grown in both fermentations. For the cultivation of fungal mycelia for large-scale production of chitosan, solid waste from solid substrate fermentation and liquid waste from submerged fermentation should be considered. Waste disposal cost and fermentation operation cost can be reduced by production of chitosan using industrial mycelia waste.

For agriculture and medical applications little is known about the metabolism of degradation products of chitin and chitosan *in vivo* and *in vitro*. ^{13}C labeled chitosan is an essential prerequisite for study the

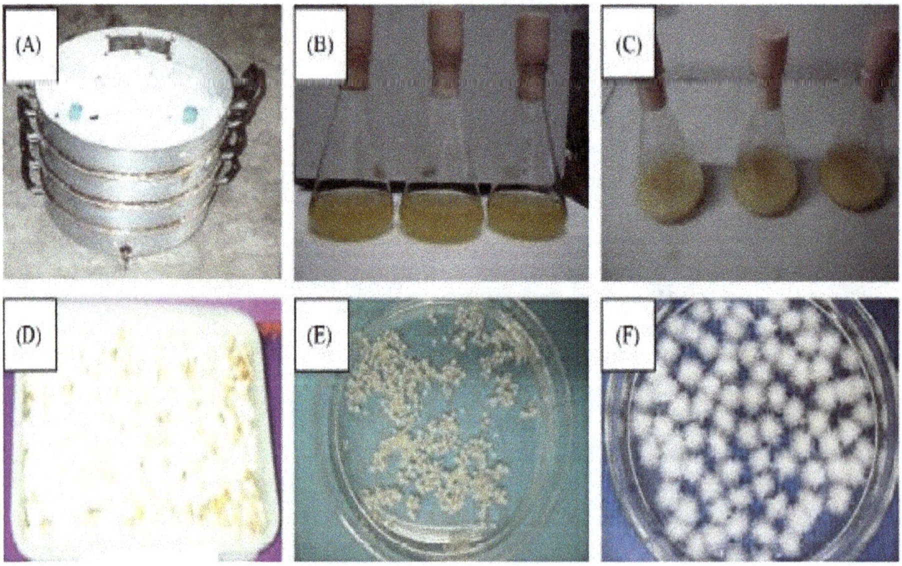
(A)
(B)
(C)
(D)
(E)
(F)

Figure 2-14 (A and D) Mycelia of fungus *Gongronella butleri* USDB 0201 grown in SSF, (B and E) mycelia of fungus *Absidia coerulea* ATCC 14076 grown in SMF, and (C and F) mycelia of fungus *Gongronella butleri* ATCC 42618 grown in SMF (Reproduced from Nwe et al., 2011, Carbohydrate Polymers, 84, 743–750. Reproduced with permission.) [10].

metabolic pathway of chito-oligosaccharides, which are degradation products of chitin and chitosan by lysozyme, chitinase and chitosanase.

For production of ^{13}C-labeled chitosan, crustacean sources could not be utilized for the biosynthesis of ^{13}C-labeled chitosan. Here cultivation of fungus in submerged fermentation medium containing ^{13}C-labeled glucose provides the fungal mycelia for production of ^{13}C-labeled chitosan. Solid substrate fermentation did not support for the production of ^{13}C-labeled chitosan from fungal mycelia [10].

The best source for the production of high molecular weight chitosan is crustacean source since the raw materials can be obtained easily in most of the countries with low cost. Fungal source is the best source for the production of low molecular weight chitosan and ^{13}C-labeled chitosan (see Table 2-1). Nowadays, commercially chitin and chitosan are produced from shells of shrimps and crabs, and bone plates of squids [6].

2-G Production of starch

Starch has been produced from various raw materials such as rice, corn, cassava roots, wheat, potato, sweet potato, etc. One of the challenges identified in starch industries is the need to develop technologies and upgrade with high level mechanisation for processing starch in order to reduce post-harvest losses [12]. Starch production process is shown in Figure 2-15.

2-H Production of DNA

DNA is abundant, inexpensive, and composed of green materials. A variety of farms waste products can be used as raw materials to produce DNA. The DNA has been produced from salmon milt and roe sacs, biowaste material left over from processing the edible parts of the fish [13]. The DNA was purified from fish waste with enzymatic process, in which 98% of the proteins are removed, decolorized with a carbon treatment, and freeze-dried [13] (Figure 2-16).

Moreover plasmid DNA also has been produced for DNA vaccine. In this production process, selection of host, vector and insert; choice of cultivation method and condition; choice of technology to harvest the biomass (filtration or centrifugation); and choice of technology for plasmid purification process (lysis method, solid separation and precipitation) are the important steps for success production of plasmid DNA (Figure 2-17) [14].

Table 2-1 Raw materials and production of chitosans from different sources.

Parameters	Fungal source			Crustacean source	Insect source		Mushroom source
	SSF		SMF				
Raw materials for chitosan production							
Source	CV	CV	IW	IW	IW	NW	IW
Variability	No	No	No	Yes	No	Yes	No
Seasonality	No	No	No	Yes	No	Yes	No
Yield of chitosan							
g of chitosan/100 g of dried raw material	6–13			4–16	–		1–4
Production of ^{13}C labeled chitosan							
Biosynthesis	Yes			No	No		Possible
****Characteristics of chitosan***							
%DD	80–90			60–90	75–85		85–90
MW (kDa)	10–450			$1–6 \times 10^3$	20–70		27–190
Purity	95%			–	–		–

(*Reproduced from Nwe et al., 2011, Biodegradable materials: production, properties and applications. Nova Science, Hauppauge, NY, Chapter 2, pp. 29–50) [6]. (SSF, solid substrate fermentation; SMF, submerged fermentation; CV, cultivation; IW, industrial waste; NW, natural waste) [6–11].

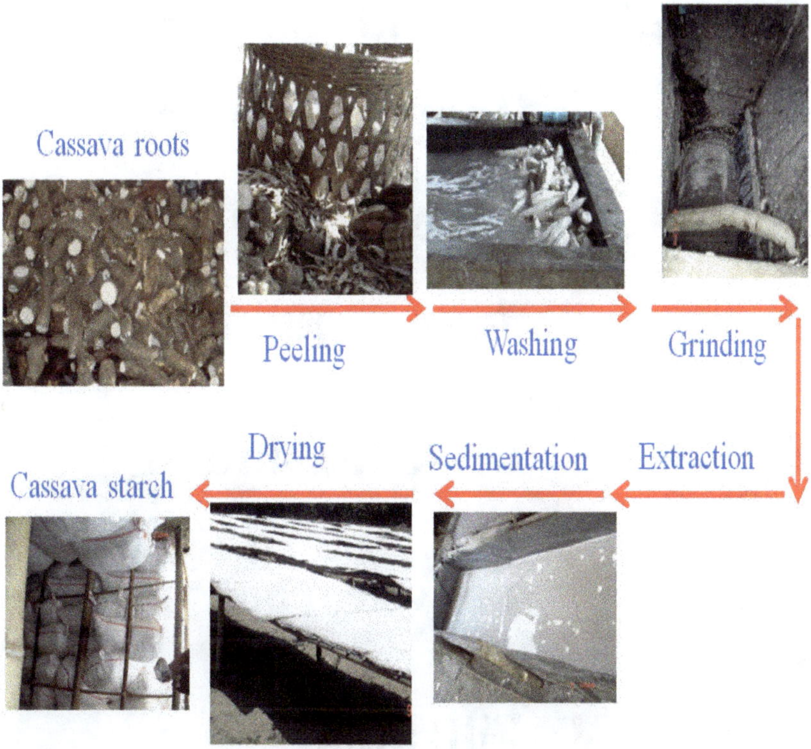
Cassava roots
Peeling
Washing
Grinding
Extraction
Sedimentation
Drying
Cassava starch

Figure 2-15 Flow chart for production of starch in industrial scale (Photos obtained from cassava starch industries in Myanmar).

2-I Production of protein

Production of protein, hemagglutinin from the avian influenza virus H_5N_1 in a baculovirus/insect cell system

Hemagglutinin protein is the receptor-binding and membrane fusion glycoprotein of influenza virus and the target for infectivity-neutralizing antibodies (Chizmadzhev, 2004, Skehel and Wiley, 2000 cited in [15]). The entire hemagglutinin protein (HA) from the H5N1 is composed of 568 amino acids, with a molecular weight of 56 kDa. The HA molecule consists of HA1 and HA2 subunits, with the HA1 subunit mediating initial contact with the cell membrane and HA2 being responsible for membrane fusion (Chizmadzhev, 2004 cited in [15]). Expression of hemagglutinin (HA1) from avian influenza H5N1 in monolayer or suspension culture insect cells by infection with the recombinant baculovirus is shown in Figure 2-18.

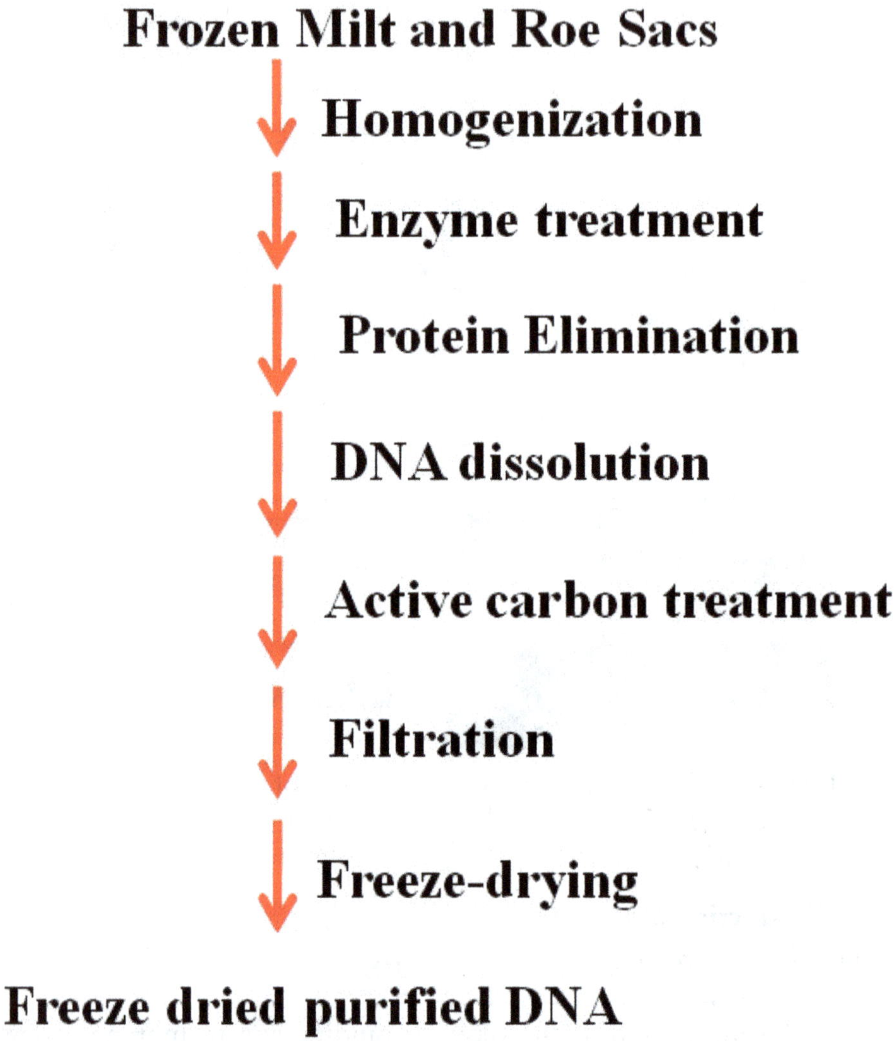
Frozen Milt and Roe Sacs
Homogenization
Enzyme treatment
Protein Elimination
DNA dissolution
Active carbon treatment
Filtration
Freeze-drying
Freeze dried purified DNA

Figure 2-16 Flow chart for production of DNA [13].

Purification of α-lactalbumin from milk

Milk contains several proteins, including caseins, β-lactoglobulin, α-lactalbumin, albumin, and immunoglobulins. Purification procedure of α-lactalbumin from milk sample is shown in Figures 2-19 and 2-20. Firstly whey solution was extracted from milk sample and then the whey sample was concentrated. IDA agarose column was set up to gel filtration apparatus and air bubbles was removed from the tube by running

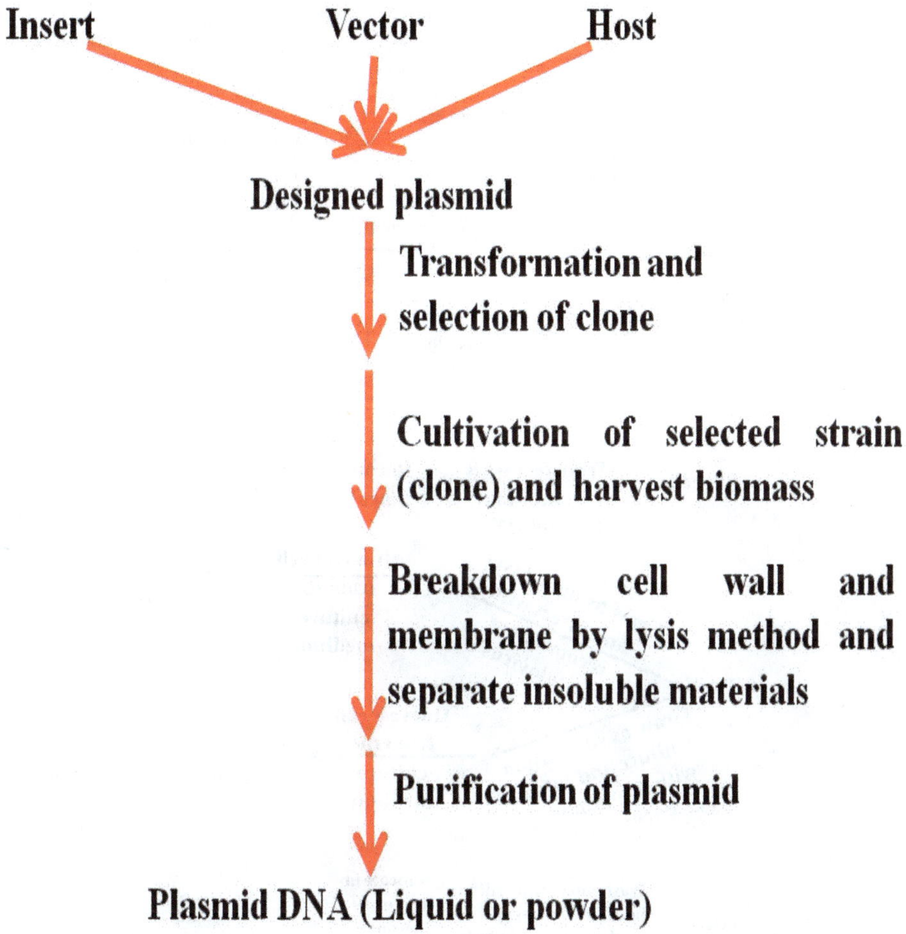
Insert
Vector
Host
Designed plasmid
Transformation and selection of clone
Cultivation of selected strain (clone) and harvest biomass
Breakdown cell wall and membrane by lysis method and separate insoluble materials
Purification of plasmid
Plasmid DNA (Liquid or powder)

Figure 2-17 Flow chart for production of plasmid DNA [14].

buffer A (0.020 M Tris, 0.5 M NaCl, pH 7.0) at maximum rate. The column was equilibrated or washed with buffer A at a flow rate of 1.5 ml/min. $CuSO_4$ solution (0.5 ml, 0.1 M) was added to the column. The column was washed with buffer A to elute all excess Cu (II). The concentrated whey, 0.5 ml was loaded to the column and the 1.5 ml fraction was collected in individual test tubes and then read A_{280} (Buffer A as blank). Buffer B was flowed at the same flow rate and then fractions were collected continuously to analyze A_{280}. The fraction of the highest reading after elution with Buffer B (0.020 M Tris, 0.5 M) was collected.

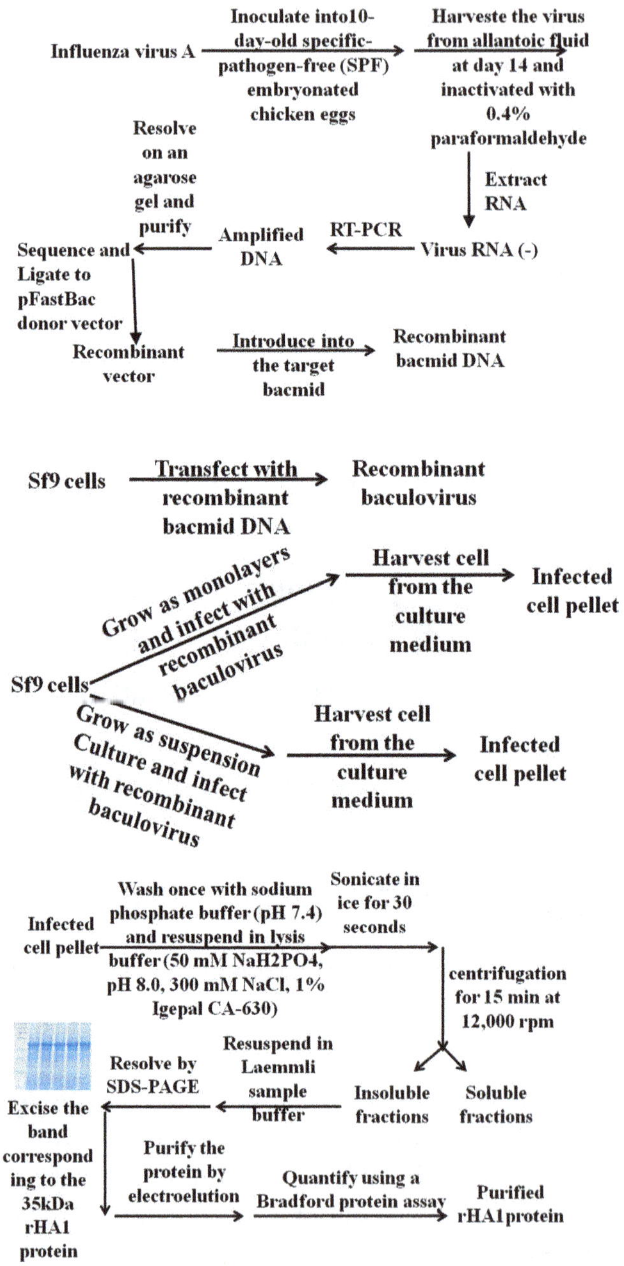
Influenza virus A
Inoculate into10-day-old specific-pathogen-free (SPF) embryonated chicken eggs
Harveste the virus from allantoic fluid at day 14 and inactivated with 0.4% paraformaldehyde
Extract RNA
Virus RNA (-)
RT-PCR
Amplified DNA
Resolve on an agarose gel and purify
Sequence and Ligate to pFastBac donor vector
Recombinant vector
Introduce into the target bacmid
Recombinant bacmid DNA
Sf9 cells
Transfect with recombinant bacmid DNA
Recombinant baculovirus
Sf9 cells
Grow as monolayers and infect with recombinant baculovirus
Harvest cell from the culture medium
Infected cell pellet
Grow as suspension Culture and infect with recombinant baculovirus
Harvest cell from the culture medium
Infected cell pellet
Infected cell pellet
Wash once with sodium phosphate buffer (pH 7.4) and resuspend in lysis buffer (50 mM NaH2PO4, pH 8.0, 300 mM NaCl, 1% Igepal CA-630)
Sonicate in ice for 30 seconds
centrifugation for 15 min at 12,000 rpm
Insoluble fractions
Soluble fractions
Resuspend in Laemmli sample buffer
Resolve by SDS-PAGE
Excise the band correspond ing to the 35kDa rHA1 protein
Purify the protein by electroelution
Quantify using a Bradford protein assay
Purified rHA1protein

Figure 2-18 Flow chart for production of purified protein in insect cells culture by infection with the recombinant baculovirus [15].

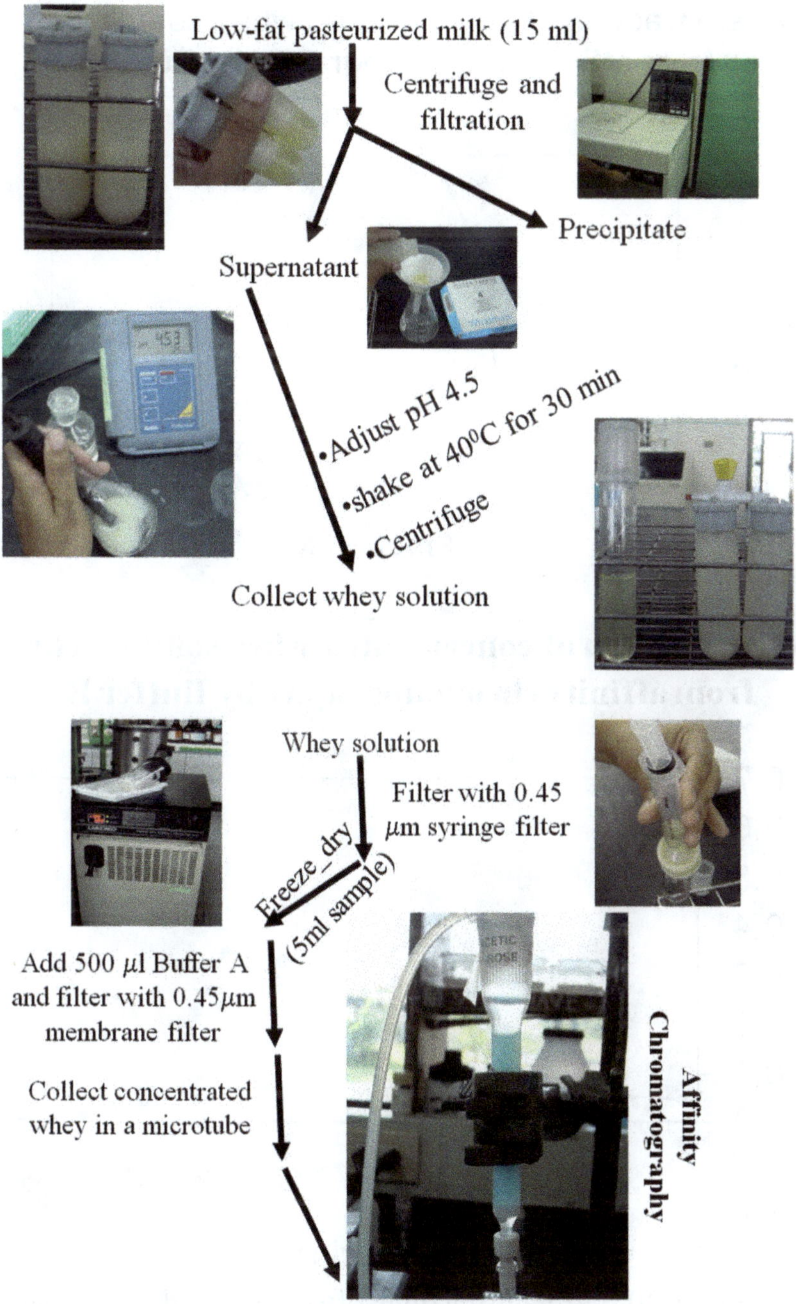
Low-fat pasteurized milk (15 ml)
Centrifuge and filtration
Precipitate
Supernatant
•Adjust pH 4.5
•shake at 40⁰C for 30 min
•Centrifuge
Collect whey solution
Whey solution
Filter with 0.45 μm syringe filter
Freeze_dry (5ml sample)
Add 500 μl Buffer A and filter with 0.45μm membrane filter
Collect concentrated whey in a microtube
Affinity Chromatography

Figure 2-19 Flow charts for extraction of α-lactalbumin from milk sample.

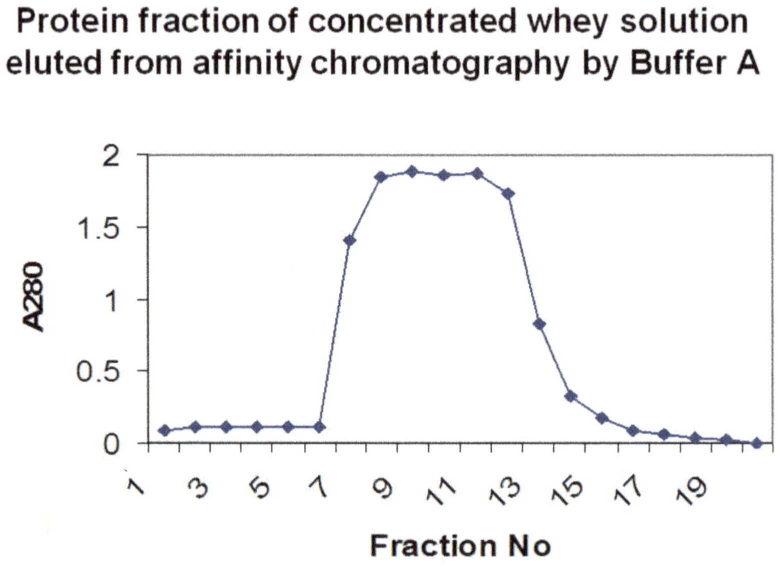
Protein fraction of concentrated whey solution eluted from affinity chromatography by Buffer A
A280
2
1.5
1
0.5
0
1
3
5
7
9
11
13
15
17
19
Fraction No

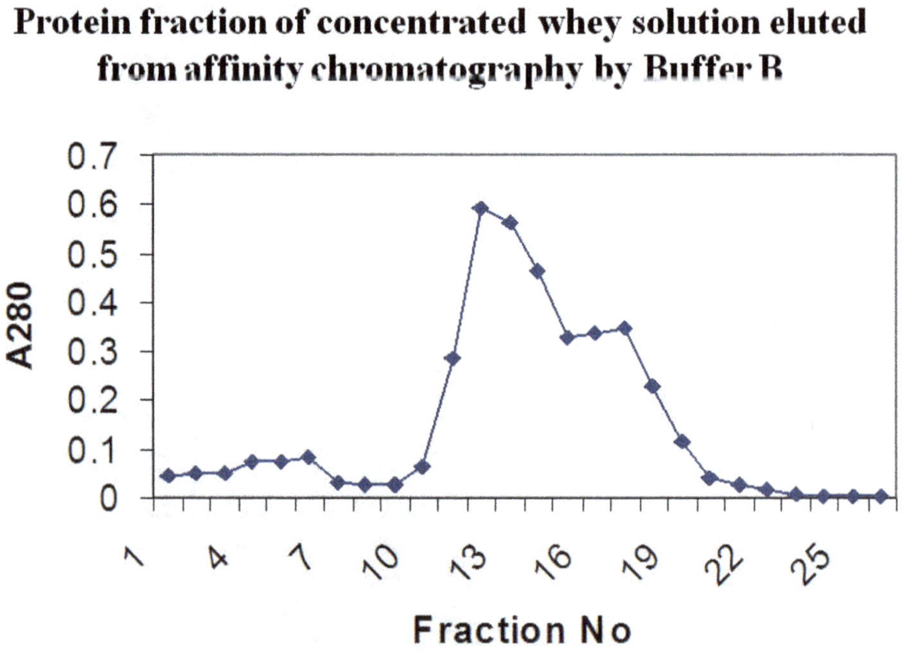
Protein fraction of concentrated whey solution eluted from affinity chromatography by Buffer B
A280
0.7
0.6
0.5
0.4
0.3
0.2
0.1
0
1
4
7
10
13
16
19
22
25
Fraction No

Figure 2-20 Protein fractions of concentrated whey solution eluted from affinity chromatography by buffer A and B.

Enzyme production

All proteins are not enzyme, but all enzymes are proteins that facilitate biochemistry reaction. The enzyme activity is highly depend on temperature, pH, concentration of enzyme and substrate, and the present of inhibitors or activators [9]. Enzymes are divided into six main groups (oxidoreductases, transferases, hydrolases, lyases, isomerases, and ligases). Enzymes have been produced from different sources (Table 2-2).

Table 2-2 Production of enzymes from different sources.

Enzyme	Production method		Substrate	Important parameters
	Source	Method		
[16]Chitin deacetylase	*Rhizopus oryzae*	Submerged fermentation	Deacetylated chitin	Fungal strain, fermentation type, fermentation medium composition and harvesting time
[17]Proteases	Plant		–	–
[17]Serine proteases, fungal proteases, endo and exo peptidases	Fermentation technology and genetic engineering		–	–

3. Characteristics of biopolymers

3-A Molecular compositions of biopolymers

The molecular compositions and structures of most of biopolymers have been determined using elemental analysis, FTIR and NMR spectrometry [1, 9]. The molecular compositions of biopolymers are shown in Table 3-1. According to their molecular compositions, the solubility properties of these biopolymers are also different from each other (Table 3-2).

Table 3-1 Molecular compositions of biopolymers.

Biopolymers	Molecular composition
Agar[1, 18]	A linear polymer of 3-1inked-β-D-galactopyranose and 4-linked 3,6-anhydro-α-L-galactopyranose
Agarose[3]	A linear polymer of alternating (1,3)-linked-β-D-galactopyranose and 1,4-linked-3, 6-anhydro-α-L-galactopyranose
Alginic acid[18]	A linear polymer based on two monomeric units, β-D mannuronic acid and α-L guluronic acid
Carrageenan[18, 19]	A 1inear polysaccharide with a repeating structure of alternating 1,3-linked β-D galactopyranose and 1,4-1inked α-D galactopyranose units. The 3-1inked units occur as the 2-and 4-sulfate or unsulfated, while the 4-1inked units occur as the 2-sulfate, 2,6-disulfate, the 3,6 anhydrid and the 3,6 anhydrid 2-sulfate Hydrocolloid composed of α-D-1,3 and β-D-1,4 galactose residues that are sulfated at up to 40% of the total weight; strong negative charge over normal pH range; associated with ammonium, calcium, magnesium, potassium, or sodium salts
Cellulose[20]	Repeat unit of two anhydroglucose rings joined via a β-1,4 glycosidic linkage

(*Continued*)

Table 3-1 Molecular compositions of biopolymers. (*Continued*)

Biopolymers	Molecular composition
Bacterial cellulose[5]	A homopolymer consisting of glucose glycosidically linked in a β-1_4 conformation
Chitin[6]	Copolymers of *N*-acetyl-D-glucosamine and D-glucosamine units linked with β-(1-4)-glycosidic bonds (D-glucosamine units less than 60%)
Chitosan[6]	Copolymers of *N*-acetyl-D-glucosamine and D-glucosamine units linked with β-(1-4)-glycosidic bonds (D-glucosamine units higher than 60%)
Starch[21]	Composed of two kinds of polysaccharides, amylose (D-glucose residues with α-(1-4) linkages) and amylopectin (D-glucose residues with (1-4) linkages and ~5% α-(1-6) branch linkages)
DNA[22]	Polynucleotide chains deoxyribonucleic acid
Protein	Polypeptides (peptides consisting of more than 50 amino acids)

Table 3-2 Solubility properties of biopolymers.

Biopolymers	Solubility
Agar	Soluble in hot water at 100°C and insoluble in cold water
Agarose	Soluble in hot water at 100°C and insoluble in cold water
Alginate	Soluble in cold water
Carrageenan[19]	λ is readily soluble in cold or hot aqueous solution; κ is soluble in hot solution; treatment of aqueous solution with potassium ion precipitates κ-carrageenan
Cellulose[23, 24, 25]	Soluble in LiCl/N,N-dimethylacetamide, DMAc, quaternary ammonium fluorides/DMSO and, ionic liquids

(*Continued*)

Table 3-2 Solubility properties of biopolymers. (*Continued*)

Biopolymers	Solubility
Bacterial cellulose[5, 26, 27]	Soluble in zinc chloride aqueous solutions, DMA/LiCl solvent system, concentrated acids: sulphuric, hydrochloric and nitric acid, 8.5% NaOH solution (solubility of cellulose increased by adding 1% of urea to the solution)
Chitin[6, 11]	Soluble in 8% (w/v)NaOH/4% (w/v) Urea, Dimethylacetamide-5% LiCl, N-methyl pyrrolidone-5% LiCl, $CaCl_2 \cdot 2H_2O$-methanol, 85% phosphoric acid, 37% HCl acid, concentrated acids (HCl, H_2SO_4, formic, acetic, dichloroacetic and trichloroacetic acid), (dimethylformamide)-N_2O_4 mixtures, hexafluoro-2-propanol, hexafluoroacetone
Chitosan[6, 9]	Soluble in dilute inorganic acids: HCl, HBr, HI, HNO_3, $HClO_4$, in concentrated H_2SO_4, in organic acids: citric, formic, lactic, acetic and pyruvic acid, in tetrahydrofuran, in ethyl acetate and in 1,2-dichloroethane, in 4% citric acid, 25% formic acid, 1% lactic acid and 1% acetic acid, 0.6% HCl and 96% H_2SO_4 and insoluble in 0.2% benzoic acid, 0.2% cinnamic acid and 6.3% oxalic acid, 3% HCl and 5% H_2SO_4 solution
Starch[28, 29]	Soluble at 80°C in ionic liquids such as 1-butyl-3-methylimidazolium chloride (BMIMCl) and 1-butyl-3-methylimidazolium dicyanamide (BMIMdca) in concentrations up to 10% (w/w), at 118°C in choline chloride/urea soluble and soluble in other non-conventional solvents such as choline chloride/oxalic acid and choline chloride/$ZnCl_2$
DNA	Soluble in water and insoluble in ethanol
Protein[30]	Denatured protein soluble in polar, protic solvents (formic acid, trifluoroacetic acid, 3-mercaptopropionic acid); some proteins dissolve in water and some protein are insoluble in water

3-B Solubility properties of biopolymers

The solubility properties of biopolymers are one of the important factors for preparation of scaffold, membrane and micro- and nano-materials. Solubility properties of biopolymers are shown in Table 3-2. The type, pattern, molecular compositions and molecular weight of polymers determine the solubility and gelling properties of biopolymers. Molecular weight and other properties of biopolymers are described in Table 3-3. Analytical methods for determination of quality and quantity of biopolymers are shown in Table 3-4.

Table 3-3 Physico-chemical properties of biopolymers.

Biopolymers	Physico-chemical properties
Agar[1, 31]	Color: White; forms: powder; Mwt: 30,000 to 100,000 Da; Gelling temperature: 35–45°C; Melting temperature: 76–100°C
Agarose[32]	Color: White; forms: powder; Mwt: 80–140 kDa; Gelling temperature: 32–45°C; Melting temperature: 80–95°C
Alginate[4, 33]	Color: White; forms: powder; Mwt: 120–311 kDa; Gelling temperature: 40°C, Melting temperature: 78–82°C
Carrageenan[19]	Color: White; forms: powder; Mwt: 20–800 kDa; Gelling temperature: lower than 40°C; Melting temperature: 40–70°C
Cellulose[5]	Color: White; forms: powder, fiber and pulp form; Mwt: approx. 142.73 kDa
Bacterial cellulose[5]	Color: White; forms: powder; Degradation temperature: >300°C
Chitin[34]	Color: White to pale yellow; forms: powder and flake form; Mwt: 8.78–10.24 × 10^6Da; Melting temperature: >300°C
Chitosan[6]	Color: White to pale yellow; forms: powder and flake form; Mwt: 10–6,000 kDa

(Continued)

Table 3-3 Physico-chemical properties of biopolymers. (*Continued*)

Biopolymers	Physico-chemical properties
Starch[35, 36, 37]	Color: White; forms: powder; Mwt: 2–7 × 10^8 Da; Gelatinisation temperatures: 69.18–81.5°C
DNA[13, 38]	Color: White to pale brown; forms: powder; Mwt: 2.5–20,000 × 10^4Da;
Protein[39]	Color: White to pale green; forms: powder; Mwt: 1.7–375 kDa

Table 3-4 Analytical methods for determination of quality and quantity of biopolymers.

Parameters	Methods
Moisture and ash content[6]	AOAC method
Turbidity[6]	Turbidity meter
Viscosity[6, 37]	Brookfield viscometer, shear dilution viscometer (CUSDC-ll series, size 150, Cannon-Ubbelohde)
Nitrogen content	Micro-Kjeldahl method, elementary analysis
Elemental analysis[1, 6]	X-ray diffraction, differential scanning calorimetry, FT-IR, near IR, solid/liquid state ^{13}C-NMR and liquid state ^{1}H-NMR, solid state ^{15}N CP/MAS NMR, 1st derivative UV spectrophotometry method, acid hydrolysis-HPLC method, acid-base titration method, elemental analysis, gel permeation chromatography, enzymatic hydrolysis-HPLC, stoichiometric dye adsorption assays, ninhydrin assay, inverse gas chromatography, acid hydrolysis-GC
Thermal decomposition of biopolymers[6]	Thermogravimetry (TG) and differential scanning calorimetry (DSC)
Molecular weight of biopolymers[1, 4, 6, 31, 32, 36, 37, 39]	Gel permeation chromatography, light scattering, viscosimetry size-exclusion chromatography, Polyacrylamide gel electrophoresis

4. Preparation of macro-, micro- and nano-materials using biopolymers

4-A Preparation of macro-, micro- and nano-sized particles using biopolymers

Preparation of chitosan coated alginate beads

Flow chart for preparation of chitosan coated Ca-alginate beads is shown in Figure 4-1. As shown in SEM images, pore size in and on the beads depend on material coated on beads.

Preparation of nanoparticles

Over the past 30 years, four methods (ionotropic gelation, microemulsion, emulsification solvent diffusion and polyelectrolyte complex) for preparation of chitosan nanoparticles have been developed based on chitosan microparticles technology [40]. Most of nanoparticles are prepared by ionotropic gelation method which is dropping a solution to biopolymer gel solution under stirring condition (Figure 4-2, Table 4-1) [40–43]. Aggregation of nanoparticles typically occurs due to the surface charge of nanoparticles (typically net positive) [44]. Bioprotectants (trehalose, mannitol and polyethylene-glycol) are used to prevent particle aggregation and to reduce mechanical stress during freezing and drying processes [43]. Thus, the most common approach to prevent aggregation is to reduce the surface charge of the nanoparticles by introducing bioprotectants [44, 45].

4-B Preparation of micro- and nano-sized fibers using biopolymers

Preparation of alginate and chitosan fibers

Alginate fibers and chitosan coated alginate fibers have been prepared using alginate, chitosan and calcium chloride solution [46]. Similar spinning system and chitosan, calcium chloride and methanol solution have been used to fabricate chitosan fibers [47]. Flow chart for spinning of chitosan fibers is shown in Figure 4-3.

Electrospinning of nanofibers

Nanofibers have been fabricated using electrospinning system [48]. Flow chart for fabrication of nanofibers is shown in Figure 4-4.

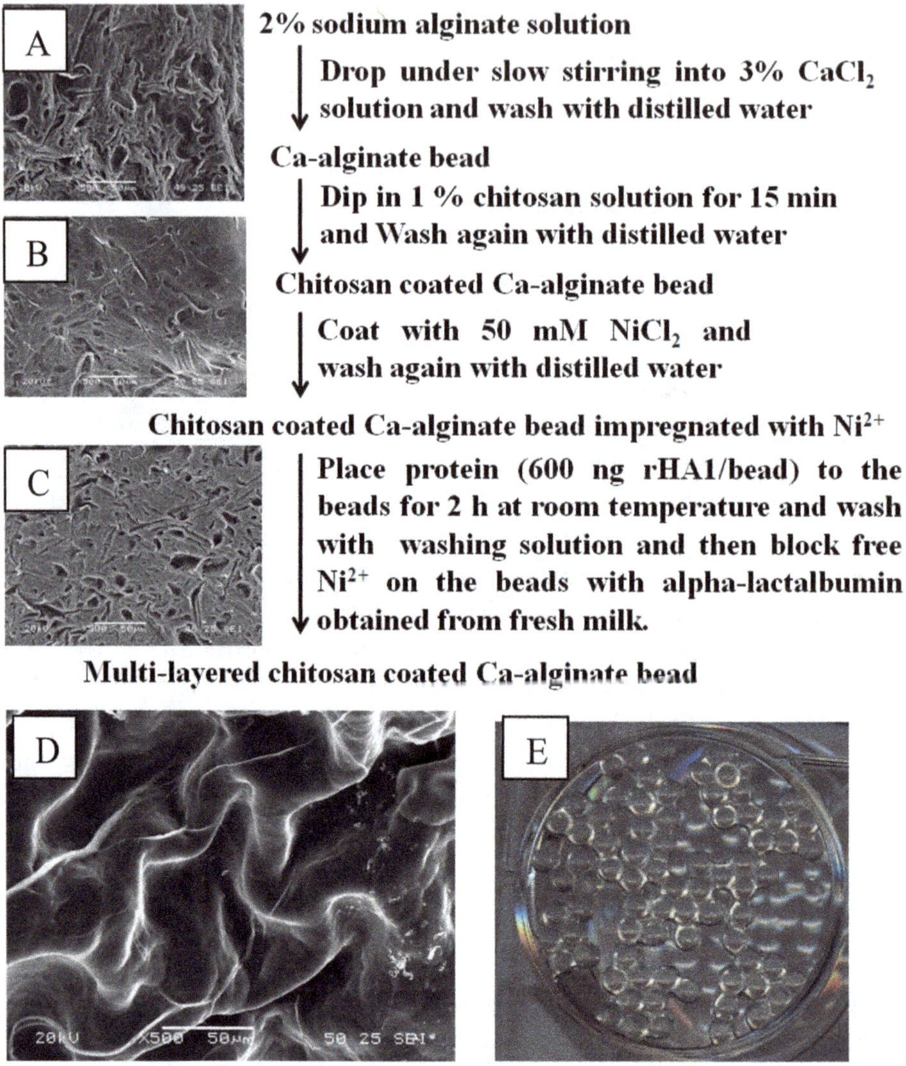
A
2% sodium alginate solution
Drop under slow stirring into 3% $CaCl_2$ solution and wash with distilled water
Ca-alginate bead
B
Dip in 1 % chitosan solution for 15 min and Wash again with distilled water
Chitosan coated Ca-alginate bead
Coat with 50 mM $NiCl_2$ and wash again with distilled water
Chitosan coated Ca-alginate bead impregnated with Ni^{2+}
C
Place protein (600 ng rHA1/bead) to the beads for 2 h at room temperature and wash with washing solution and then block free Ni^{2+} on the beads with alpha-lactalbumin obtained from fresh milk.
Multi-layered chitosan coated Ca-alginate bead
D
E

Figure 4-1 Flow chart for preparation of multi-layered chitosan coated Ca-alginate beads (E) and SEM images of bead in the preparation process (A: Ca-alginate bead, B: chitosan coated Ca-alginate bead, C: Chitosan coated Ca-alginate bead impregnated with Ni^{2+}, D: Multi-layered chitosan coated Ca-alginate bead).

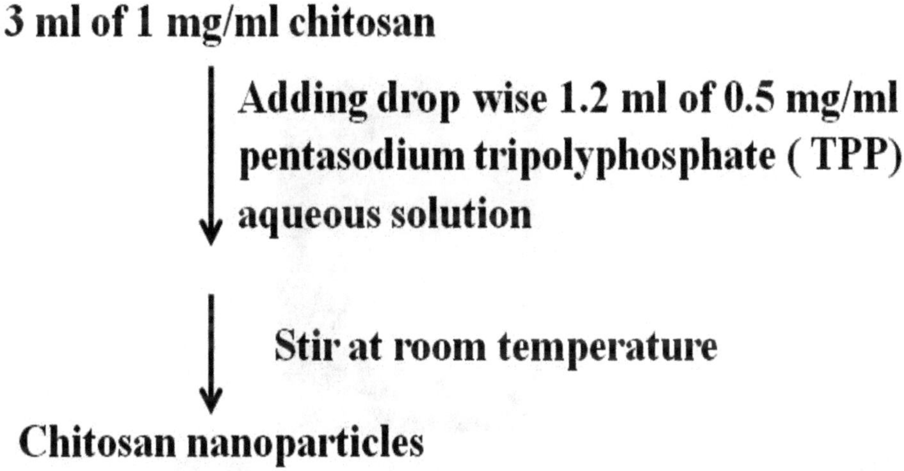
3 ml of 1 mg/ml chitosan
Adding drop wise 1.2 ml of 0.5 mg/ml pentasodium tripolyphosphate (TPP) aqueous solution
Stir at room temperature
Chitosan nanoparticles

Figure 4-2 Flow chart for preparation of chitosan nanoparticles [42, 43].

Table 4-1 Preparation of biopolymer nanoparticle.

Type of nanoparticles	Ionic solution	Polymer gel solution	Bioprotectants
Chitosan nanoparticles[42, 43]	Tripolyphosphate-pentasodium solution	Chitosan solution	Trehalose, mannitol and polyethylene-glycol
siRNA nanoparticles[44, 45]	Biotin-tagged scrambled siRNA and tripolyphosphate-pentasodium solution	Peptide-tagged polyethylene glycol (PEG) ylated chitosan solution	Poly(ethylene glycol) (PEG), or sugar molecules (e.g., cyclodextrin), and hyaluronic acid
Sodium alginate nanoparticles[41]	Poly-L-lysine	Calcium chloride and sodium alginate solution	–

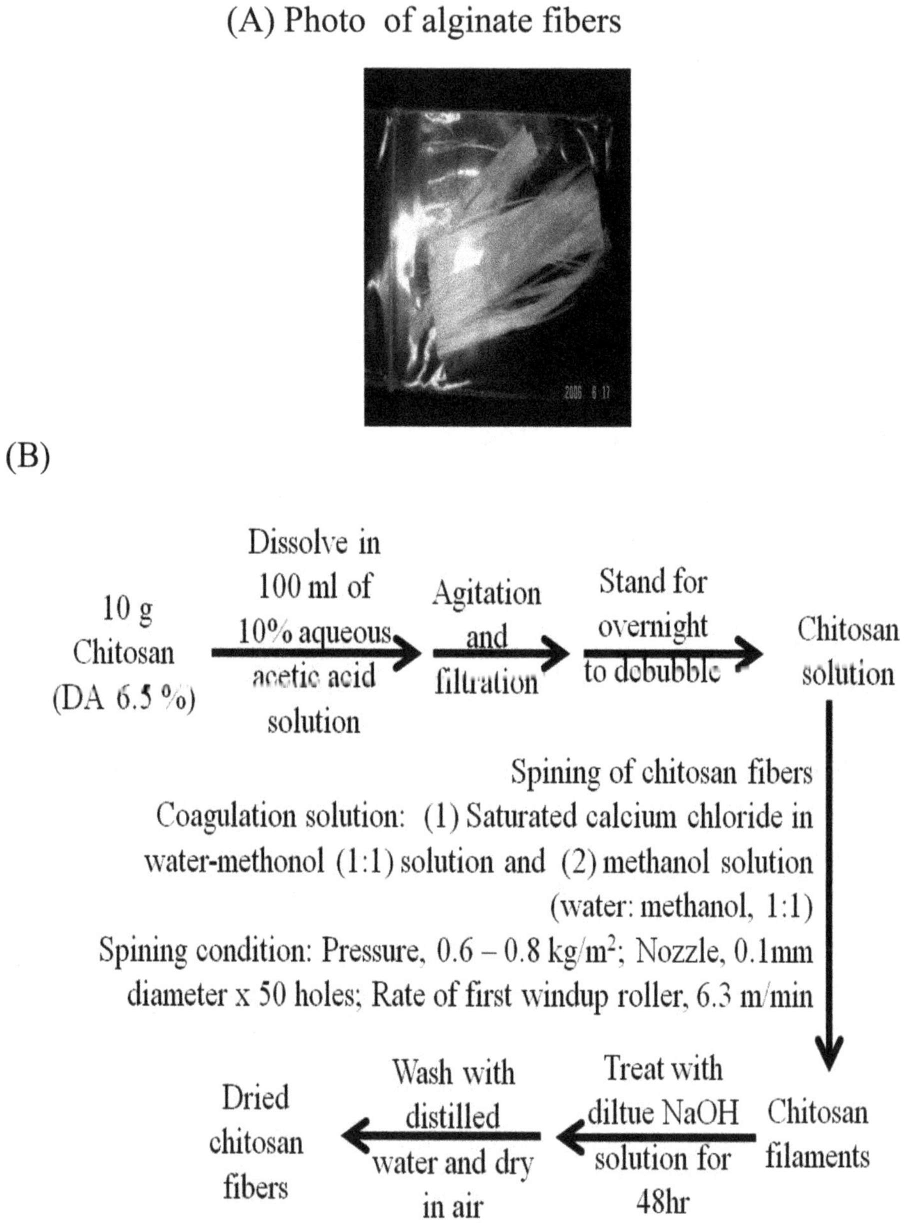
(A) Photo of alginate fibers
(B)
10 g Chitosan (DA 6.5 %)
Dissolve in 100 ml of 10% aqueous acetic acid solution
Agitation and filtration
Stand for overnight to debubble
Chitosan solution
Spining of chitosan fibers
Coagulation solution: (1) Saturated calcium chloride in water-methonol (1:1) solution and (2) methanol solution (water: methanol, 1:1)
Spining condition: Pressure, 0.6 – 0.8 kg/m²; Nozzle, 0.1mm diameter x 50 holes; Rate of first windup roller, 6.3 m/min
Chitosan filaments
Treat with diltue NaOH solution for 48hr
Wash with distilled water and dry in air
Dried chitosan fibers

Figure 4-3 (A) Photo of alginate fibers (B) Flow chart for preparation of chitosan fibers [47].

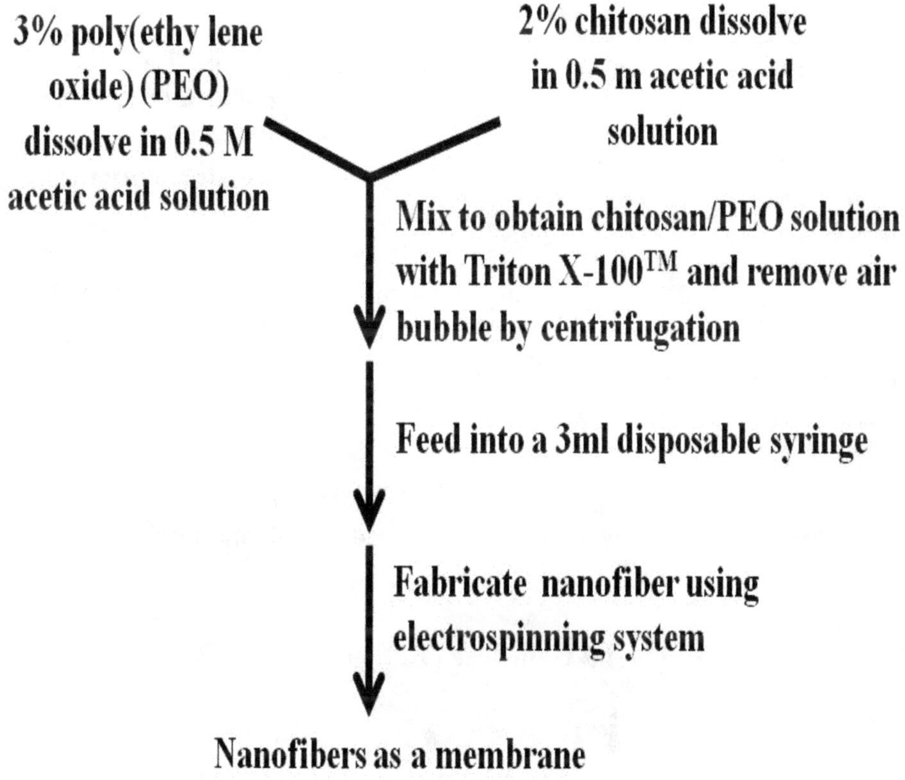
3% poly(ethy lene oxide) (PEO) dissolve in 0.5 M acetic acid solution
2% chitosan dissolve in 0.5 m acetic acid solution
Mix to obtain chitosan/PEO solution with Triton X-100TM and remove air bubble by centrifugation
Feed into a 3ml disposable syringe
Fabricate nanofiber using electrospinning system
Nanofibers as a membrane

Figure 4-4 Flow chart for fabrication of nanofibers [48].

4-C Preparation of micro- and nano-sized pores in membranes using biopolymers

Preparation of alginate membrane crosslinked with Ca^{2+}

Firstly alginate membrane was prepared using alginate solution [49]. After that the membrane was placed in calcium chloride solution to crosslink alginate with Ca^{2+} (Figure 4-5).

Preparation of bacterial cellulose (BC) membrane

The BC gel (Figure 4-6) was prepared and BC membrane was prepared according to the procedure shown in Figure 4-6 [49].

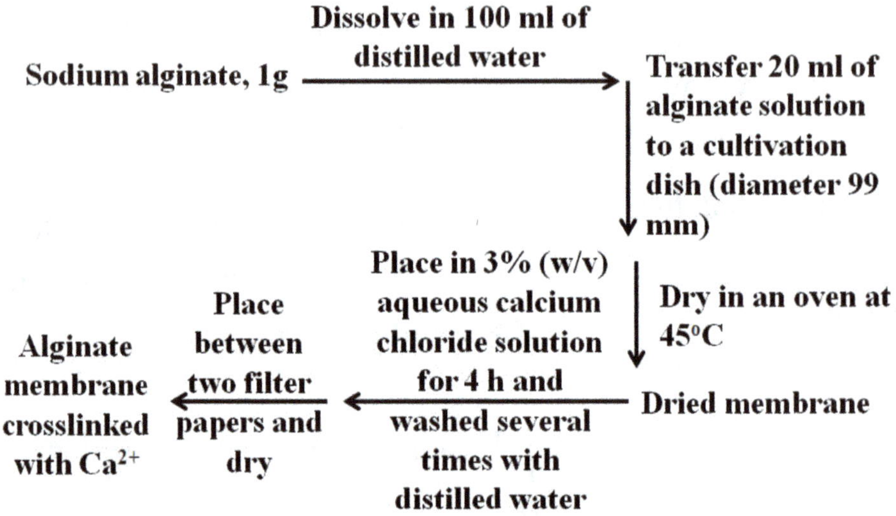
Sodium alginate, 1g
Dissolve in 100 ml of distilled water
Transfer 20 ml of alginate solution to a cultivation dish (diameter 99 mm)
Dry in an oven at 45°C
Dried membrane
Place in 3% (w/v) aqueous calcium chloride solution for 4 h and washed several times with distilled water
Place between two filter papers and dry
Alginate membrane crosslinked with Ca2+

Figure 4-5 Flow chart for preparation of alginate membrane crosslinked with Ca^{2+} [49].

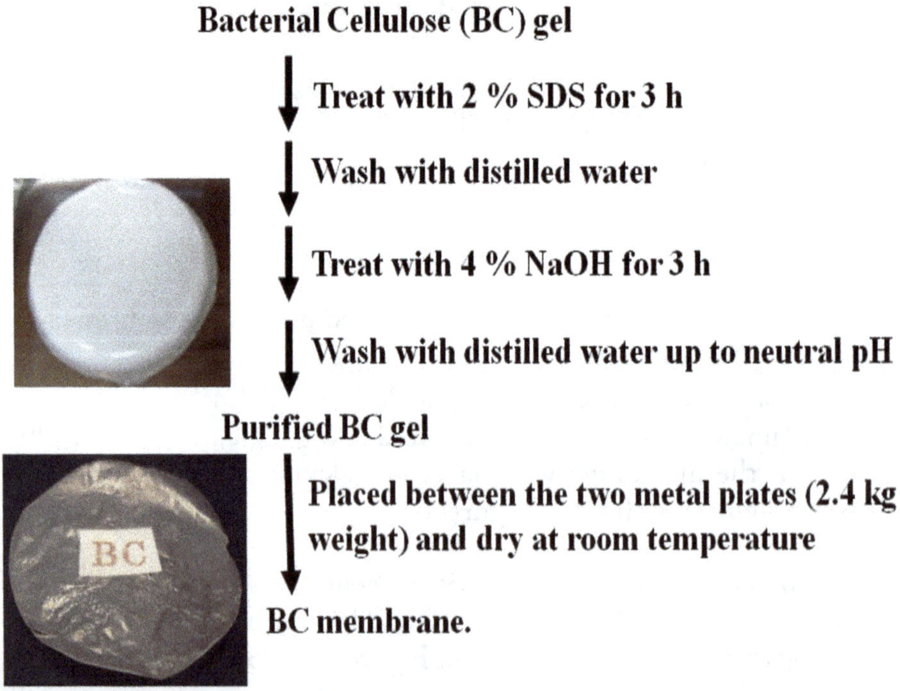
Bacterial Cellulose (BC) gel
Treat with 2 % SDS for 3 h
Wash with distilled water
Treat with 4 % NaOH for 3 h
Wash with distilled water up to neutral pH
Purified BC gel
Placed between the two metal plates (2.4 kg weight) and dry at room temperature
BC membrane.
BC

Figure 4-6 Flow chart for preparation of bacterial cellulose membrane [49].

Preparation of glutaraldehyde (GTA) crosslinked gelatin membrane

Membranes have been prepared using cow bone gelatin, fish skin gelatin or pork skin gelatin crosslinked with glutaradehyde [49]. The properties of membranes depend on ratio of gelatin/GTA. The procedure for preparation of glutaraldehyde crosslinked gelatin membrane is shown in Figure 4-7 [49].

Preparation of chitosan membrane

Chitosan membrane has been prepared using chitosan solution dissolved in acetic acid and solvent casting method (Figure 4-8) [11, 50]. The neutralized membrane was placed between two filter papers and dried at room temperature [11]. Moreover macroporous and microporous chitosan membranes have been prepared using silica particles and polyethylene glycol respectively. Silica particles and polyethylene glycol in chitosan membranes dissolved in sodium hydroxide solution and porous structure constructed in the membranes (Figure 4-9) [50].

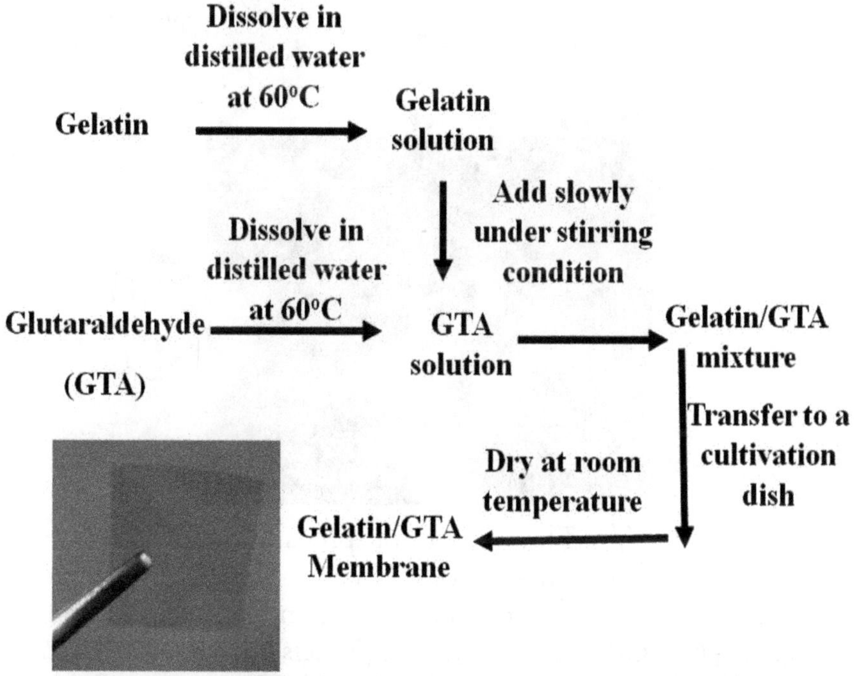
Dissolve in distilled water at 60°C
Gelatin
Gelatin solution
Add slowly under stirring condition
Dissolve in distilled water at 60°C
Glutaraldehyde (GTA)
GTA solution
Gelatin/GTA mixture
Transfer to a cultivation dish
Dry at room temperature
Gelatin/GTA Membrane

Figure 4-7 Flow chart for preparation of glutaraldehyde crosslinked gelatin membrane [49].

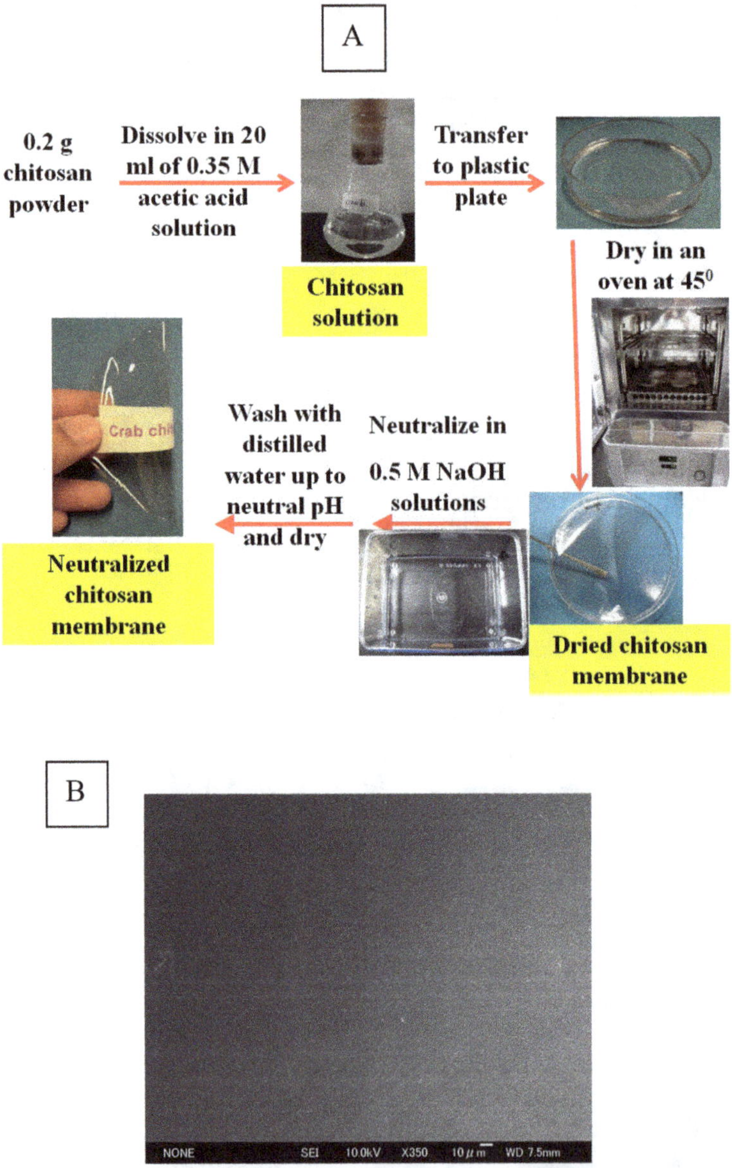
A
0.2 g chitosan powder
Dissolve in 20 ml of 0.35 M acetic acid solution
Chitosan solution
Transfer to plastic plate
Dry in an oven at 45⁰
Dried chitosan membrane
Neutralize in 0.5 M NaOH solutions
Wash with distilled water up to neutral pH and dry
Neutralized chitosan membrane
Crab chi
B
NONE SEI 10.0kV X350 10 μm WD 7.5mm

Figure 4-8 (A) Flow chart for preparation of chitosan membrane (Reproduced from Nwe et al., 2013, Marine Biomaterials: Characterization, Isolation and Application, (pp. 45–60). CRC press, Taylor & Francis Group, New York. Reproduced with permission.) [11]; (B) SEM image of chitosan membrane.

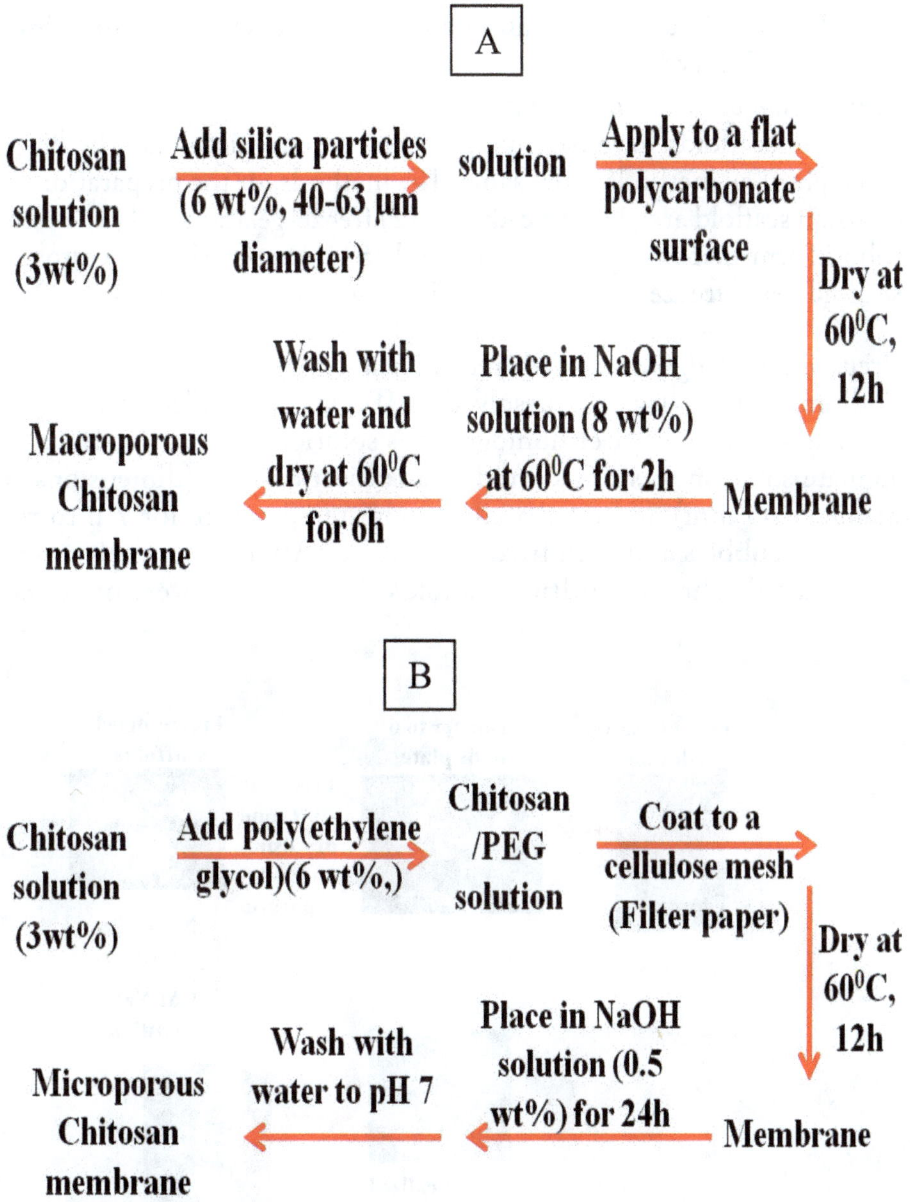
A
Chitosan solution (3wt%)
Add silica particles (6 wt%, 40-63 µm diameter)
solution
Apply to a flat polycarbonate surface
Dry at 60⁰C, 12h
Membrane
Place in NaOH solution (8 wt%) at 60⁰C for 2h
Wash with water and dry at 60⁰C for 6h
Macroporous Chitosan membrane
B
Chitosan solution (3wt%)
Add poly(ethylene glycol)(6 wt%,)
Chitosan /PEG solution
Coat to a cellulose mesh (Filter paper)
Dry at 60⁰C, 12h
Membrane
Place in NaOH solution (0.5 wt%) for 24h
Wash with water to pH 7
Microporous Chitosan membrane

Figure 4-9 (A) Flow chart for preparation of macroporous chitosan membrane, and (B) Microporous chitosan membrane [50].

4-D Preparation of micro- and nano-sized pores in scaffolds using biopolymers

Preparation of chitosan scaffolds

Chitosan scaffolds have been prepared by several methods. The basic principle of all methods is the same. The methods for the preparation of chitosan scaffold are (1) freeze-drying (2) freeze-gelation and (3) 3-axis robotic arm dispensing system method [51]. Preparation of chitosan scaffold using freeze-drying method is shown in Figure 4-10.

Preparation of alginate scaffold crosslinked with Ca^{2+}

Sodium alginate, 1 g was dissolved in 100 ml of distilled water with the help of a mixer. After homogeneous solution was obtained, 8 g of alginate solution was transferred to a cultivation dish (dimensions = diameter 35 mm) and incubated at room temperature for 7 h to remove air bubbles and then frozen at –20 °C. After that samples were freeze-dried. The freeze-dried scaffolds (Figure 4-11) were dissolved

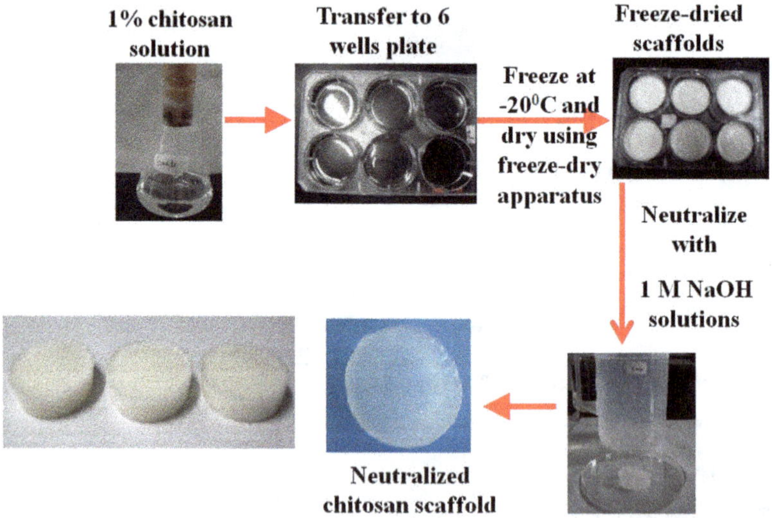
1% chitosan solution
Transfer to 6 wells plate
Freeze-dried scaffolds
Freeze at -20°C and dry using freeze-dry apparatus
Neutralize with
1 M NaOH solutions
Neutralized chitosan scaffold

Figure 4-10 Preparation of chitosan scaffold using freezing-drying method (Reproduced from Nwe et al., 2013, Marine Biomaterials: Characterization, Isolation and Application, (pp. 45–60). CRC press, Taylor & Francis Group, New York. Reproduced with permission.) [11].

rapidly in a neutral aqueous medium. In order to prevent the scaffold dissolution, the scaffolds were placed in 3.6% (w/v) aqueous calcium chloride solutions dissolved in water for overnight and washed several times with distilled water to remove excess calcium chloride in and on the scaffolds.

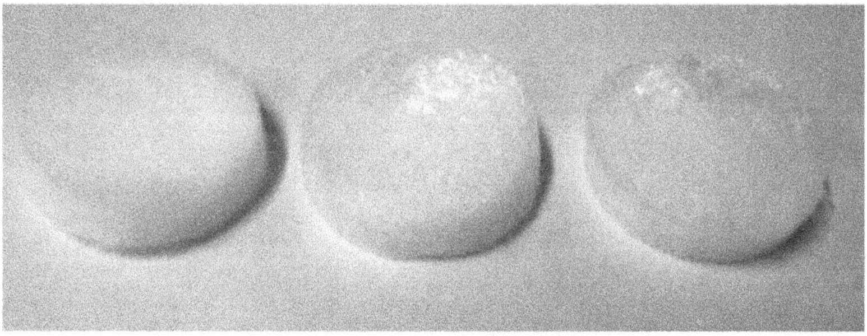

Figure 4-11 Freeze-dried alginate scaffolds.

5. Characterization of macro-, micro- and nano-biomaterials

The pore size, porosity, mechanical strength, sensitivity to lysozyme degradation, degree of swelling, water vapour permeability and proliferation of cells on matrixes are important parameters to use biomaterials in various applications [6, 52].

5-A Morphology, pore size and porosity of micro- and nano-biomaterials

Morphology and pore size of biomaterials have been studied using scanning electron microscope. In these analyses samples were mounted on sample stubs and coated with an ultrathin layer of palladium in a coating apparatus and then the pore morphology of materials was observed under a scanning electron microscope (JEOL-JSM 6700, Japan) at an accelerating voltage of 5 kV and current of 10 μA. At the same time, the diameter of pores in each sample was measured using the JEOL-Image software. The total number of pores analyzed for each sample was more than 10. The values were expressed as the mean value of independent replicates [53].

The porosity of scaffolds has been determined using different methods:

- I: Porosity (%) = (ρ material − ρ sponge)/ρ material × 100 [54];
- II: Microcomputerized tomography analysis [55–57];
- III: Scanning electron microscopy [56];
- IV: Porosity = (Pore volume of scaffold/Total volume of scaffold) × 100 [58];
- V: Volume percent of porogen in a scaffold [57];
- VI: Mercury intrusion porosimetry [57]. There are advantages and disadvantages coupling in each method. In which the biological properties of biomaterials such as biodegradation and biocompatibility were studied on water absorbed biomaterials. The determination of the porosity of biomaterials in dry form and in wet form might be different, and then for the application of biomaterials in tissue engineering, cells were cultured on the biomaterials in wet state. Therefore the method for determination of porosity in biomaterials should

ideally be performed in wet state. Therefore a method to determine porosity of biomaterials of water absorbed biomaterials was developed in my research [53]. In this analysis dried biomaterials was soaked overnight in water. Three pieces of tissue paper were dried overnight in an oven at 55°C and then the weight of tissue papers (W1) was measured. The water absorbed biomaterials was taken from the water medium and the diameter and thickness of the biomaterials were measured and the volume of the water absorbed materials, V1 = ($\pi r^2 h$) then calculated. The water absorbed materials was placed on the top of the tissue papers in a centrifuge tube and then centrifuged at 6000 rpm for 5 min. After that biomaterials was removed from the tissue papers and the weight of wet tissue papers (W2) was measured and then the weight of water in the void space of biomaterials, W3 = (W2 – W1) was calculated using Equation (5-1). The volume of water in the void space, V2 was determined by dividing the weight of water in the void space of biomaterials, W3 by the density of water (1.0).

$$\text{Porosity of water absorbed biomaterials (\%)} = (V2/V1) \times 100 \qquad (5\text{-}1)$$

5-B Water absorption property of macro-, micro- and nano-biomaterials

Water absorption properties of biomateria have been determined with the ratio of weight of absorbed water to weight of dried biomateria. Dried biomateria with a known weight (Wi) was soaked overnight in phosphate buffered saline (PBS buffer, 8 g NaCl, 0.2 g KCl, 1.44 g Na_2HPO_4 and 0.24 g KH_2PO_4 per liter, pH 7.4, 7 ml) or distilled water at room temperature. After that sample was removed from soaking medium, the wet weight of the material (Ww) was measured. For membrane samples, the sample was blotted on filter paper to remove excess water from the membrane and then the wet weight of the material (Ww) was measured. The weight of water absorbed in material (Wa) was calculated as a fraction of dry weight of the material as shown in Equation (5-2). The values were expressed as the means value of three independent replicates [49, 53].

$$\text{Wa (g water/g of scaffold)} = (\text{Ww} - \text{Wi})/\text{Wi} \qquad (5\text{-}2)$$

5-C Mechanical properties of micro- and nano-biomaterials

Neutralized materials were cut into strips in wet or dry condition to prevent deformation of the samples. For the wet sample, the materials was freeze-dried after by soaking in water. The dried sample was sandwiched between two pieces of sandpaper on either side of material with the help of Scotch permanent adhesive glue. About 2.5 mm of the material strip was glued on either side to the end of the paper pieces. The tensile properties of materials were examined with an Orientec StA-1150 (Japan) at room temperature. Sand papers were attached to both sides of the grips. The end of each grip was the beginning of the attached sample in order to prevent the deformation of samples during mounting. The applied force and elongation at break were determined. The values are expressed as the mean of independent replicates [53].

5-D *In vitro* biodegradation of micro- and nano-biomaterials

The biodegradation rate of biomaterials was determined *in vitro* by measuring the change in sample weight over time under treatment with lysozyme under specific conditions. The initial dry weight of materials was determined and recorded as W1. The materials were sterilized in PBS buffer at 121°C for 15 min. After cooling, free solution was removed from the sterilized biomaterials and incubated in 7 mL of sterilized PBS buffer containing 10 μg/mL hen egg white (HEW) lysozyme. All incubations were done in six well plates at 37°C in a humidified 5% CO_2 environmental incubator. Media were replaced weekly with freshly prepared lysozyme solution. After 14 days of incubation, the samples were removed from the degradation media, washed with distilled water and freeze-dried. The weight of the freeze-dried materials was recorded as W2. The percentage degradation weight of the materials was calculated using Equation (5-3). The values were expressed as the mean value of three independent replicates [53].

$$\text{Degradation weight (\%)} = (W1 - W2)/W1 \times 100 \qquad (5\text{-}3)$$

5-E Determination of metal content in micro- and nano-biomaterials

Metal content in biomaterials has been determined using sequential plasma emission spectrometer or atomic absorption spectroscopy. To determine calcium content in biomaterials, firstly all glassware were washed with a soap solution, rinsed with deionized water, soaked in 1M HCl solution, washed with deionized water and dried. The membrane was dried in an oven at 65°C for 24 h and weighted and then placed into a test tube (dimensions = 18 mm × 180 mm). Nitric acid, 61% (4 ml) and 2 ml of hydrogen peroxide were added to the sample tube and then heated to 110°C for 3 h. After cooling, the digested sample was transferred to 100 ml volumetric flasks and made the sample volume up to 100 ml with deionized water. Calcium content in the digested sample was analyzed by sequential plasma emission spectrometer (ICPS-7510, Shimadzu, Japan). Analytical standards were prepared using calcium standard solutions (Wako Pure Chemical Industries, Ltd., Osaka, Japan). Data analysis was performed using Microsoft Excel [49].

5-F Attachment, morphology, viability and proliferation of life cells on micro- and nano-biomaterials

For the determination of attachment and morphology of fibroblast NIH/3T3 cells, the dried samples were cut into small pieces and placed in test vials containing 2 ml of distilled water and then sterilized in an autoclave at 121°C for 15 min. After cooling, free water was removed from the sterilized samples and each sample was seeded with 1 ml of cell suspension (8.2×10^4 cells/ml) and then incubated in a humidified 5% CO_2 incubator for 4 h at 37°C. After 4 h incubation, the cell-seeded sample was washed with 1 ml of DMEM medium for 3 times and collected the washing media and then centrifuged at 1000 rpm for 5 min to precipitate the unattached cells. The precipitated cells were collected by decanting the washing media and the cell pellets were re-suspended in each 100–200 ml DMEM medium. The number of unattached cells in the suspension was counted with a hemocytometer and the volume of the cell suspensions was measured with the help of micropipettes. The number of attached cells on the samples was estimated by subtraction the number of unattached cells from initial number of cells in the inoculum. The cell-seeded samples were incubated in each 5 ml of DMEM medium at 37°C in a humidified 5% CO_2 incubator. Fresh medium was provided every 1–3 days depending on the incubation period. The number of proliferated cells was counted after 5–6 days of incubation. Prior to cell counting, sample was washed 3 times with PBS buffer to remove unattached cells and FBS from the medium. The sample was immersed in 2 ml of 1× trypsin–EDTA aqueous solution and then incubated at 37°C for 10 min to detach the cells from the sample. The cell suspension was then sheared mildly with a micropipette to detach the cells from the sample material. After detachment of all cells from the sample (confirmed by microscopic observation), the sample was removed from the cell suspension and the cell suspension was centrifuged at 1000 rpm for 5 min. The precipitated cells were resuspensed in 100–200 ml of DMEM medium and counted the number of proliferated cells using a hemacytometer. To assess the cell viability and morphology, the sterilized sample were seeded with 150–300 ml fibroblast NIH/3T3 cell suspension (6–10 × 10^4 cells/ml) and then incubated at 37°C in a humidified 5% CO_2 incubator. After 2 and 6 days incubation, cell cultured samples were washed 3 times with PBS buffer and incubated in 1 ml of 2 mg fluorescein diacetate/ml solution (FDA, Wako Pure Chemicals, Japan) dissolved in PBS buffer at 37°C for 10 min to stain viable cells green and then viewed under a confocal laser scanning microscope (Carl Zeiss Laser Scanning Microscopy, Axiovert 200 M, LSM5PASCAL, Germany) [49, 53].

Methods and machines to determine the characteristics of micro- and nano-materials are described in Table 5-1. The observation of

Table 5-1 Analytical methods for determination of characteristics of micro- and nano-biomaterials.

Parameters	Methods and machines
Structure of the nanoparticles in the polymer matrix	X-ray diffraction patterns[59, 60]
Morphology of the films	SEM[49, 59]
Thickness of films	Hand-held micrometer (dial thickness gauge 7301, Mitutoyo)[59]
Color and transparency films	CR-300 Minolta Chroma Meter (Minolta Camera Co., Ltd., Osaka, Japan)[59]
Contact angle of water of films	Contact angle analyzer (model Phoenix 150, Surface Electro Optics Co. Ltd., Kunpo, Korea)[59]
Water solubility	The percentage of dry matter solubilized in distilled water[49, 59]
Size of nano-particles	SEM and TEM[60]

cross-sections of nano-materials can be performed by various types of equipments including transmission electron microscope (TEM), scanning tunneling microscope (STM) and atomic force microscope (AFM), as well as scanning electron microscope (SEM).

6. Applications of macro-, micro- and nano-biomaterials prepared using biopolymers

6-A Application of micro- and nano-biomaterials prepared using agar

The wide range of gel properties makes agar suitable for a diversity of uses in bacteriological, biotechnological, medical, pharmaceutical, industrial and food applications (Table 6-1) [1, 31, 61, 62]. Microorganisms and plant do not use agar as nutrient and agar can make various forms.

6-B Application of micro- and nano-biomaterials prepared using agarose

Agarose is widely used not only in the field of biotechnology but also in many industries [3]. In biotechnological applications, agarose gels are widely used for separation of biological molecules (Table 6-2).

6-C Application of micro- and nano-biomaterials prepared using alginate

The uses of alginate are based on soluble in water and on various forms products. Alginate has been used as thickeners, stabilizers, gelling agents and emulsifiers in the food, textile, paper making and pharmaceutical industries (Table 6-3).

6-D Application of micro- and nano-biomaterials prepared using carrageenan

Carrageenan has been used in milk products, processed meats, dietetic formulations, infant formula, toothpaste, cosmetics, skin preparations, pesticides, and laxatives (Table 6-4). Moreover carrageenan is a suitable support material for the immobilisation of whole cells, as proven by several applications in different industrial processes.

6-E Application of micro- and nano-biomaterials prepared using plant and bacterial cellulose

Cellulose materials of various origins are built with nanofibrils [67]. Moreover nanocellulose has been produced using various methods and this material has been used in various applications [67]. Various applications of cellulose are shown in Table 6-5.

Table 6-1 Agar in food, bacteriological and biotechnological applications.

Forms	Applications
Powder	Bacteriology and biotechnology
Gel[2, 31]	Food (cake, jelly, cheese, yoghurt)
Nanoparticles[61, 62]	Antimicrobial matrix

Table 6-2 Various applications of agarose.

Forms	Applications
Microcapsules[63]	Tissue engineering, cell therapy and biopharmaceutical applications
Gel[3, 15]	Gel filtration, ion exchange chromatography, electrophoresis, immobilisation of biocatalysts
Agarose hydrogel nanoparticles[64]	Protein and peptide drug delivery

Table 6-3 Various applications of alginate.

Forms	Applications
Gel[2]	Textile printing, food (ice cream, salad dressing, mayonnaise, yoghurts, longer lasting beer foam, clarification of wine, jellies), paper making, binder for fish feeds
Films[2]	Preserve frozen fish
Bead[2]	Immobilised biocatalysts, controlled release of drug
Fibers[2]	Wound dressing
Nanoparticles[65]	Controlled release of drug

Table 6-4 Various applications of carrageenan.

Forms	Applications
Beads[2, 19]	Immobilised cells for food industries (vinegar, fermented milk, beer, ethanol), for pharmaceutical applications (tetracycline, chlorotetracycline, antibiotics, D-aspartic acid), for cleaning of industrial effluents (nitrogen removal, 4-chlorophenol degradation, 2,4,6-trichlorophenol degradation), and for pentachlorophenol pesticide degradation in contaminant soil
Powder or gel[2, 19]	Used as emulsifier, stabilizer, colloid or gum in food products (soymilk, chocolate and other flavoured milks, dairy products, infant formulas, nutritional supplement beverages, cottage cheese), used as binder in toothpaste, used as air freshener gel
Nanocomposite superabsorbents[66]	Remove cationic crystal violet dye from water

Table 6-5 Various applications of cellulose.

Forms	Applications
Bacterial cellulose	
Gel[5, 49]	Food applications (nata de coco), tissue engineering, wound dressing
Membrane[5, 49]	Tissue engineering, cosmetic and medical
Nanocomposites[20]	Electroluminescent layer
Plant cellulose	
Fiber[5]	Textile industry
Nano-cellulose particles and nano-fibrillated cellulose[20, 67]	Paper making, reinforcing adhesives, foams and aerogels, nanopaper and starch–nanocellulose composites

6-F Application of micro- and nano-biomaterials prepared using chitin and chitosan

Chitosan and chitin are attractive compounds to be used in medical and biological products. Various applications of chitin and chitosan are shown in Table 6-6.

6-G Application of micro- and nano-biomaterials prepared using starch

Starches have been used in food, textile, medical and pharmaceutical, syrup and brewing, cosmetic, and paper and glues industries. Various applications of starch are shown in Table 6-7.

6-H Applications of micro- and nano-biomaterials prepared using DNA

Biomaterials prepared using DNA show great promise for numerous applications [13]. Applications of biomaterials prepared using DNA are shown in Table 6-8.

6-I Application of micro- and nano-biomaterials prepared using proteins

There are different types of proteins. Some proteins are soluble in water and some proteins are insoluble in water. Based on their solubility properties, proteins have been used to prepare various forms of materials. These biomaterials have been applied in many sectors such as agriculture, food, feed, biotechnology, pharmaceutical, etc. Applications of biomaterials prepared using proteins are shown in Table 6-9.

Table 6-6 Various applications of chitin and chitosan.

Forms	Applications
Solution[6]	Flocculating agent
Nonoparticle[6, 40, 42]	Lead ion adsorbent, drug delivery system
Bead[6]	Zinc removal agent (sorption of heavy metal ions)
Powder[6]	Antioxidative agent
Scaffold[6, 53]	Hemostatic agent, tissue regeneration template
Membrane[6]	Tissue regeneration template, edible film, semipermeable membrane
Nanopowder[6]	Yogurt supplement

Table 6-7 Various applications of starch.

Forms	Applications
Hydrogel membrane[68]	Wound dressing
Nano powder[69]	Electrode
Powder[70–72]	Food industries (noodles, sweeteners, dextrins, monosodium glutamate (MSG), modified starch, crackers, cakes, syrups, desserts, puddings, creams, bread, cookies, varies fresh snacks); paper industries (coatings and sizing in paper); textile and carpets industries (binders, adhesives, absorbents); brewing industries; glue industries; cosmetics industries; medicine pharmaceuticals industries; ceramics and construction (spray concrete)
Nano crystals[73]	Nanoscale biofiller
Scaffold[74]	Tissue engineering

Table 6-8 Various applications of DNA.

Forms	Applications
Thin films[13]	Optical, electronic, and electro-optic applications
Nanoparticles[13, 75, 76]	Electronics applications, gene therapy, and sensors
Carbon nanotube-doped DNA[13]	Electronics applications

Table 6-9 Various applications of proteins.

Forms	Protein	Applications
Solution[15]	Hemagglutinin	Inactivated vaccine
Powder[17]	Protein hydrolysates	Animal cell culture (monoclonal antibodies, therapeutic, proteins); animal feeds; agriculture (fertilizers); food (flavor enhancers)
Powder and solution	Enzymes	Catalytic, food analysis
Membrane[49]	Gelatin	Biodegradable tissue regeneration template
Nanoparticles[77]	Albumin and gelatin	Drug delivery carriers

7. Conclusion

In this monograph, production of biopolymers, properties of biopolymers, preparation of macro-, micro- and nano-materials using biopolymers, characterization of biomaterials and application of biomaterials are described. Various form of biomaterials such as powder, solution, scaffolds, gel, films, beads, fibers, microcapsules, nano-particles, etc have been used in many sectors such as food production, biotechnology, waste water treatment, electronic applications, biomedical and nano-medical technologies, etc. The cumulative information given in this monograph is hoped to serve researchers, who are interested in biomedicine, nano-medicine, biopolymers and various fields of biotechnology. Moreover it is hoped that these information will serve a great scope of further research in these fields for better applications and will be useful to companies interested in production and applications of biopolymers.

References

[1] Villanueva, R. D., Sousa, A. M. M., Gonçalves, M. P., Nilsson, M., and Hilliou, L., 2010, "Production and Properties of Agar from the Invasive Marine Alga, *Gracilaria vermiculophylla* (Gracilariales, Rhodophyta)," *Journal of Applied Phycology*, 22, pp. 211–220.

[2] McHugh, D. J., 2003, "A Guide to the Seaweed Industry," *FAO fisheries technical paper*, No. 441, Rome, pp. 20–72.

[3] Jeon, Y., Athukorala, Y., and Lee, J., 2005, "Characterization of Agarose Product from Agar Using DMSO," *Algae*, 20, pp. 61–67.

[4] Saude, N., Chèze-Lange, H., Beunard, D., Dhulster, P., Guillochon, D., Cazé, A. M., Morcellet, M., and Junter, G. A., 2002, "Alginate Production by *Azotobacter vinelandii* in a Membrane Bioreactor," *Process Biochemistry*, 38, pp. 273–278.

[5] Chawla, P. R., Ishwar, B., Bajaj, I. B., Survase, S. A., and Singhal, R. S., 2009, "Microbial Cellulose: Fermentative Production and Applications," *Food Technology and Biotechnology*, 47, pp. 107–124.

[6] Nwe, N., Furuike, T., and Tamura, H., 2011, "Chitosan from Aquatic and Terrestrial Organisms and Microorganisms: Production, Properties and Applications," *Biodegradable Materials: Production, Properties and Applications*, B. M. Johnson, Z. E. Berkel, eds., Nova Science, Hauppauge, NY, Chapter 2, pp. 29–50.

[7] Nwe, N., Furuike, T., and Tamura, H., 2011, "Chitin and Chitosan from Terrestrial Organisms," *Chitin, Chitosan, Oligosaccharides and Their Derivatives: Biological Activities and Applications*, S. K. Kim, ed., CRC press, Taylor & Francis Group, New York, pp. 3–10.

[8] Nwe, N., Furuike, T., and Tamura, H., 2011, "Production, Properties and Applications of Fungal Cell Wall Polysaccharides: Chitosan and Glucan," *Advances in Polymer Science*, 244, pp. 187–208.

[9] Nwe, N., Furuike, T., and Tamura, H., 2010, "Production of Fungal Chitosan by Enzymatic Method and Applications in Plant Tissue Culture and Tissue Engineering: 11 years of Our Progress, Present Situation and Future Prospects," *Biopolymers*, M. Elnashar, ed., Rijeka, SCIYO, pp. 135–162.

[10] Nwe, N., Furuike, T., Osaka, I., Fujimori, H., Kawasaki, H., Arakawa, R., Tokura, S., Stevens, W. F., Kurozumi, S., Takamori, Y., Fukuda, M., and Tamura, H., 2011, "Laboratory Scale Production of ^{13}C labeled Chitosan by Fungi *Absidia coerulea* and *Gongronella butleri* Grown in Solid Substrate and Submerged Fermentation," *Carbohydrate Polymers*, 84, pp. 743–750.

[11] Nwe, N., Furuike, T., and Tamura, H., 2013, "Isolation and Characterization of Chitin and Chitosan as Potential Biomaterials," *Marine Biomaterials: Characterization, Isolation and Applications*, S. K. Kim, ed., CRC press, Taylor & Francis Group, New York, pp. 45–60.

[12] Olukunle, O. J., and Olukunle, O. F., 2007, "Development of a Sustainable System for Cassava Starch Extraction," *Conference on International Agricultural Research for Development*, University of Göttingen, October 9–11.

[13] Grote, J., 2008, "Biopolymer Materials Show Promise for Electronics and Photonics Applications," *Proc. SPIE*, DOI 10.1117/2.1200805.1082.

[14] Prather, K. J., Sagar, S., Murphy, J., and Chartrain, M., 2003, "Industrial Scale Production of Plasmid DNA for Vaccine and Gene Therapy: Plasmid Design, Production, and Purification," *Enzyme and Microbial Technology*, 33, pp. 865–888.

[15] Nwe, N., He, Q., Damrongwatanapokin, S., Du, Q., Manopo, I., Limlamthong, Y., Fenner, B. J., Spencer, L., and Kwang, J., 2006, "Expression of Hemagglutinin Protein from the Avian Influenza Virus H5N1 in a Baculovirus/insect Cell System Significantly Enhanced by Suspension Culture," *BMC Microbiology*, 6, p. 16.

[16] Aye, K. N., Nwe, N., and Stevens, W. F., 2004, "Chitin Deacetylase Enzyme from *Rhizopas oryzae*," *Proceeding of 6th Asia Pacific Chitin and Chitosan Symposium*, May 23–26, Singapore, S4-1.

[17] Pasupuleti, V. K., and Braun, S., 2010, "State of the Art Manufacturing of Protein Hydrolysates," V. K. Pasupuleti and A. L. Demain (eds.), *Protein Hydrolysates in Biotechnology*, Chapter 2, pp. 11–32.

[18] Istini, S., Ohno, M., and Kusunose, H., 1994, "Methods of Analysis for Agar, Carrageenan and Alginate in Seaweed," *Bulletin of Marine Science and Fisheries*, Kochi Univ, 14, pp. 49–55.

[19] Necas, J., and Bartosikova, L., 2013, "Carrageenan: A Review," *Veterinarni Medicina*, 58, pp. 187–205.

[20] Eichhorn, S. J., Dufresne, A., Aranguren, M., Marcovich, N. E., Capadona, J. R., Rowan, S. J., Weder, C., Thielemans, W., Roman, M., Renneckar, S., Gindl, W., Veigel, S., Keckes, J., Yano, H., Abe, K., Nogi, M., Nakagaito, A. N., Mangalam, A., Simonsen, J., Benight, A. S., Bismarck, A., Berglund, L. A., and Peijs, T., 2010, "Review: Current International Research into Cellulose Nanofibres and Nanocomposites," *Journal of Materials Science*, 45, pp. 1–33.

[21] Robyt, J. F., 2008, "Starch: Structure, Properties, Chemistry, and Enzymology," *Glycoscience*, pp. 1437–1472.

[22] Watson, J. D., and Crick, F. H. C., 1953, "The Structure of DNA," *Cold Spring Harbor Symposia on Quantitative Biology*, 18, pp. 123–131.

[23] Ramos, L. A., Morgado, D. L., Gessner, F., Frollini, E., and Seoud, O. A. E., 2011, "A Physical Organic Chemistry Approach to Dissolution of Cellulose: Effects of Cellulose Mercerization on Its Properties and on the Kinetics of Its Decrystallization," *Organic Chemistry in Argentina*, 7, pp. 416–425.

[24] Swatloski, R. P., Spear, S. K., Holbrey, J. D., and Rogers, R. D., 2002, "Dissolution of Cellose with Ionic Liquids," *Journal of American Chemical Society*, 124, pp. 4974–4975.

[25] Olsson, C., and Westman, G., 2013, "Direct Dissolution of Cellulose: Background, Means and Applications," *Cellulose – Fundamental Aspects*, Chapter 6, pp. 143–178.

[26] Lu, X., and Shen, X., 2011, "Solubility of Bacteria Cellulose in Zinc Chloride Aqueous Solutions," *Carbohydrate Polymers*, 86, pp. 239–244.

[27] Costa, L. M. M., Olyveira, G. M., Basmaji, P., and Filho, L. X., 2012, "Nanopores Structure in Electrospun Bacterial Cellulose," *Journal of Biomaterials and Nanobiotechnology*, 3, pp. 92–96.

[28] Biswas, A., Shogren, R. L., Stevenson, D. G., Willett, J. L., and Bhowmik, P. K., 2006, "Ionic Liquids as Solvents for Biopolymers: Acylation of Starch and Zein Protein," *Carbohydrate Polymers*, 66, pp. 546–550.

[29] Zdanowicz, M., and Spychaj, T., 2011, "Ionic Liquids as Starch Plasticizers or Solvents," *Polimery*, 56, pp. 861–864.

[30] Houen, G., Svaerke, C., and Barkholt, V., 1999, "The Solubilities of Denatured Proteins in Different Organic Solvents," *Acta Chemica Scandinavica*, 53, pp. 1122–1126.

[31] Uzuhashi, Y., and Nishinari, K., 2003, "Physicschemical Properties of Agar and Its Utilization in Food and Related Industry," *Foods Food Ingredients J. Jpn*, 208, No. 10.

[32] Rochas, C., and Lahaye, M., 1989, "Average Molecular Weight and Molecular Weight Distribution of Agarose and Agarose-type Polysaccharides," *Carbohydrate Polymers*, 10, pp. 289–298.

[33] Parthiban, C., Parameswari, K., Saranya, C., Hemalatha, A., and Anantharaman, P., 2012, "Production of Sodium Alginate from Selected Seaweeds and Their Physiochemical and Biochemical Properties," *Asian Pacific Journal of Tropical Biomedicine*, pp. 1–4.

[34] Chandumpa, A., Singhpibulporn, N., Faroongsarng, D., and Sornprasit, P., 2004, "Preparation and Physico-chemical Characterization of Chitin and Chitosan from the Pens of the Squid Species, *Loligo lessoniana* and *Loligo formosana*," *Carbohydrate Polymers*, 58, pp. 467–474.

[35] Aprianita, A., Purwandari, U., Watson, B., and Vasiljevic, T., 2009, "Physico-chemical Properties of Flours and Starches from Selected Commercial Tubers Available in Australia," *International Food Research Journal*, 16, pp. 507–520.

[36] Xie, Y., Yan, M., Yuan, S., Sun, S., and Huo, Q., 2013, "Effect of Microwave Treatment on the Physicochemical Properties of Potato Starch Granules," *Chemistry Central Journal*, 7, p. 113.

[37] Millard, M. M., Dintzis, F. R., Willett, J. L., and Klavons, J. A., 1997, "Light-scattering Molecular Weights and Intrinsic Viscosities of Processed Waxy Maize Starches in 90% Dimethyl Sulfoxide and H_2O," *Carbohydrates*, 74, pp. 687–691.

[38] Blin, N., and Stafford, D. W., 1976, "A General Method for Isolation of High Molecular Weight DNA from Eukaryotes," *Nucleic Acids Research*, 3, pp. 2303–2308.

[39] Schmitz, O., Boison, G., Salzmann, H., Bothe, H., Schu, K., Wang, S., and Happe, T., 2002, "HoxE—A Subunit Specific for the Pentameric Bidirectional Hydrogenase Complex (HoxEFUYH) of Cyanobacteria," *Biochimica et Biophysica Acta*, 1554, pp. 66–74.

[40] Tiyaboonchai, W., 2003, "Chitosan Nanoparticles: A Promising System for Drug Delivery," *Naresuan University Journal*, 11, pp. 51–66.

[41] Moradhaseli, S., Mirakabadi, A. Z., Sarzaeem, A., Mohammadpour Dounighi, N., Soheily, S., and Borumand, M. R., 2013, "Preparation and Characterization of Sodium Alginate Nanoparticles Containing ICD-85 (Venom Derived Peptides)," *International Journal of Innovation and Applied Studies*, 4, No. 3 Nov., pp. 534–542.

[42] Salamanca, A. E., Diebold, Y., Calonge, M., Garcıá-Vazquez, C., Callejo, S., Vila, A., and Alonso, M. J., 2006, "Chitosan Nanoparticles as a Potential Drug Delivery System for the Ocular Surface: Toxicity, Uptake Mechanism and in Vivo Tolerance," *Investigative Ophthalmology & Visual Science*, 47, pp. 1416–1425.

[43] Rampino, A., Borgogna, M., Blasi, P., Bellich, B., and Cesàro, A., 2013, "Chitosan Nanoparticles: Preparation, Size Evolution and Stability," *International Journal of Pharmaceutics*, 455, pp. 219–228.

[44] Tokatlian, T., and Segura, T., 2010, "siRNA Applications in Nanomedicine," *Nanomed Nanobiotechnol*, 2, pp. 305–315.

[45] Malhotra, M., Tomaro-Duchesneau, C., Saha, S., and Prakash, S., 2013, "Systemic siRNA Delivery via Peptide-tagged Polymeric Nanoparticles, Targeting PLK1 Gene in a Mouse Xenograft Model of Colorectal Cancer," *International Journal of Biomaterials*, 2013, Article ID 252531, http://dx.doi.org/10.1155/2013/252531.

[46] Tamura, H., Tsurata, Y., and Tokura, S., 2002, "Preparation of Chitosan-coated Alginate Filament," *Material Science and Engineering*, C20, pp. 143–147.

[47] Tamura, H., Tsuruta, Y., Itoyama, K., Worakitkanchanakul, W., Rujiravanit, R., and Tokura, S., 2004, "Preparation of Chitosan Filament Applying New Coagulation System," *Carbohydrate Polymers*, 56, pp. 205–211.
[48] Bhattarai, N., Edmondson, D., Veiseh, O., Matsen, F. A., and Zhang, M., 2005, "Electrospun Chitosan-based Nanofibers and Their Cellular Compatibility," *Biomaterials*, 26, pp. 6176–6184.
[49] Nwe, N., Furuike, T., and Tamura, H., 2010, "Selection of a Biopolymer Based on Attachment, Morphology and Proliferation of Fibroblast NIH/3T3 Cells for the Development of a Biodegradable Tissue Regeneration Template: Alginate, Bacterial Cellulose and Gelatin," *Process Biochemistry*, 45, pp. 457–466.
[50] Clasen, C., Wilhelms, T., and Kulicke, W. M., 2006, "Formation and Characterization of Chitosan Membranes," *Biomacromolecules*, 7, pp. 3210–3222.
[51] Nwe, N., and Stevens, W. F., 2008, "Preparation and Characteristics of Chitosan Scaffolds for Tissue Engineering," In: *Recent Research in Biomedical Aspects of Chitin and Chitosan*, H. Tamura, ed., Research Signpost, ISBN 978-81-308-0254-1, India, pp. 57–68.
[52] Nwe, N., and Stevens, W. F., 2008, "Production of Chitin and Chitosan and Their Applications in the Medical and Biological Sector," In: *Recent Research in Biomedical Aspects of Chitin and Chitosan*, H. Tamura, ed., Research Signpost, ISBN 978-81-308-0254-1, India, pp. 161–176.
[53] Nwe, N., Furuike, T., and Tamura, H., 2009, "The Mechanical and Biological Properties of Chitosan Scaffolds for Tissue Regeneration Templates are Significantly Enhanced by Chitosan from *Gongronella butleri*," *Materials*, 2, pp. 374–398.
[54] Katoh, K., Tanabe, T., and Yamauchi, K., 2004, "Novel Approach to Fabricate Keratin Sponge Scaffolds with Controlled Pore Size and Porosity," *Biomaterials*, 25, pp. 4255–4262.
[55] Ghosh, S., Gutierrez, V., Fernández, C., Rodriguez-Perez, M. A., Viana, J. C., Reis, R. L., and Mano, J. F., 2008, "Dynamic Mechanical Behavior of Starch-based Scaffolds in Dry and Physiologically Simulated Conditions: Effect of Porosity and Pore Size," *Acta Biomaterialia*, 4, pp. 950–959.
[56] Kasten, P., Beyen, I., Niemeyer, P., Luginbühl, R., Bohner, M., and Richter, W., 2008, "Porosity and Pore Size of beta-tricalcium Phosphate Scaffold can Influence Protein Production and Osteogenic Differentiation of Human Mesenchymal Stem Cells: An in Vitro and in Vivo Study," *Acta Biomaterialia*, 4, pp. 1904–1915.
[57] Shi, X., Sitharaman, B., Pham, Q. P., Liang, F., Wu, K., Billups, W. E., Wilson, L. J., and Mikos, A. G., 2007, "Fabrication of Porous Ultra-short Single-walled Carbon Nanotube Nanocomposite Scaffolds for Bone Tissue Engineering," *Biomaterials*, 28, pp. 4078–4090.
[58] Oh, S. H., Park, I. K., Kim, J. M., and Lee, J. H., 2007, "In Vitro and in Vivo Characteristics of PCL Scaffolds with Pore Size Gradient Fabricated by a Centrifugation Method," *Biomaterials*, 28, pp. 1664–1671.
[59] Rhim, J. W., Hong, S. I., Park, H. M., and Ng, P. K. W., 2006, "Preparation and Characterization of Chitosan-based Nanocomposite Films with Antimicrobial Activity," *Journal of Agricultural and Food Chemistry*, 54, pp. 5814–5822.
[60] Pandey, S., Goswami, G. K., and Nanda, K. K., 2012, "Green Synthesis of Biopolymer–silver Nanoparticle Nanocomposite: An Optical Sensor for Ammonia Detection," *International Journal of Biological Macromolecules*, 51, pp. 583–589.
[61] Muthuswamy, E., Sree Ramadevi, S., Vasan, H. N., Garcia, C., Noé, L., and Verelst, M., 2007, "Highly Stable Ag Nanoparticles in Agar-agar Matrix as Inorganic–organic Hybrid," *Journal of Nanoparticle Research*, 9, pp. 561–567.

[62] Ghosh, S., Kaushik, R., Nagalakshmi, K., Hoti, S. L., Menezes, G. A., Harish, B. N., and Vasan, H. N., 2010, "Antimicrobial Activity of Highly Stable Silver Nanoparticles Embedded in Agar–agar Matrix as a Thin Film," *Carbohydrate Research*, 345, pp. 2220–2227.

[63] Sakai, S., Hashimoto, I., and Kawakami, K., 2008, "Production of Cell-enclosing Hollow-core Agarose Microcapsules via Jetting in Water-immiscible Liquid Paraffin and Formation of Embryoid Body-like Spherical Tissues from Mouse ES Cells Enclosed within These Microcapsules," *Biotechnology and Bioengineering*, 99, pp. 235–243.

[64] Wang, N., and Wu, X. S., 1997, "Preparation and Characterization of Agarose Hydrogel Nanoparticles for Protein and Peptide Drug Delivery," *Pharmaceutical Development and Technology*, 2, pp. 135–142.

[65] Asadi, A., 2013, "Streptomycin-loaded PLGA-alginate Nanoparticles: Preparation, Characterization, and Assessment," *Applied Nanoscience*, DOI 10.1007/s13204-013-0219-8.

[66] Mahdavinia, G. R., and Zhalebaghy, R., 2012, "Removal Kinetic of Cationic Dye Using Poly (Sodium Acrylate)-Carrageenan/Na-Montmorillonite Nanocomposite Superabsorbents," *Journal of Materials and Environmental Science*, 3, pp. 895–906.

[67] Ioelovich, M., 2008, "Cellulose as a Nanostructured Polymer," *Bioresources*, 3, pp. 1403–1418.

[68] Pal, K., Banthia, A. K., and Majumdar, D. K., 2006, "Preparation of Transparent Starch Based Hydrogel Membrane with Potential Application as Wound Dressing," *Trends in Biomaterials and Artificial Organs*, 20, pp. 59–67.

[69] Gaur, M. S., and Gaur, K., 2010, "Spontaneous Current Study in Wheat Starch Nano Powder," *Acta Physica Polonica A*, 117, pp. 945–948.

[70] Chen, Z., Schols, H. A., and Voragen, A. G. J., 2003, "Starch Granule Size Strongly Determines Starch Noodle Processing and Noodle Quality," *Journal of Food Science*, 68, pp. 1584–1589.

[71] Grommers, H. E., and van der Krogt, D. A., 2009, "Potato Starch: Production, Modifications and Uses," *Starch: Chemistry and Technology*, J. BeMiller, and R. Whistler, eds., Chapter 11, pp. 511–539.

[72] Vasanthan, T., and Hoover, R., 2009, "Barley Starch: Production, Properties, Modification and Uses," *Starch: Chemistry and Technology*, J. BeMiller, and R. Whistler, eds., Chapter 16, pp. 601–628.

[73] Lin, N., Huang, J., Chang, P. R., Anderson, D. P., and Yu, J., 2011, "Preparation, Modification, and Application of Starch Nanocrystals in Nanomaterials: A Review," Journal *of Nanomaterials*, 2011, Article ID 573687, doi:10.1155/2011/573687.

[74] Salgado, A. J., Coutinho, O. P., and Reis, R. L., 2004, "Novel Starch-based Scaffolds for Bone Tissue Engineering: Cytotoxicity, Cell Culture, and Protein Expression," *Tissue Engineering*, 10, pp. 465–74.

[75] Shi, Q., Tiera, M. J., Zhang, X., Dai, K., Benderdour, M., and Fernandes, J. C., 2011, "Chitosan-DNA/siRNA Nanoparticles for Gene Therapy," *Non-Viral Gene Therapy*, X. Yuan, ed., ISBN: 978-953-307-538-9, http://www.intechopen.com/books/non-viral-genetherapy/chitosan-dna-sirna-nanoparticles-for-gene-therapy.

[76] Heuer-Jungemann, A., Harimech, P. K., Brown, T., and Kanaras, A. G., 2013, "Gold Nanoparticles and Fluorescently-labelled DNA as a Platform for Biological Sensing," *Nanoscale*, 5, pp. 9503–9510.

[77] Lohcharoenkal, W., Wang, L., Chen, Y. C., and Rojanasakul, Y., 2014, "Protein Nanoparticles as Drug Delivery Carriers for Cancer Therapy," *BioMed Research International*, 2014, Article ID 180549, http://dx.doi.org/10.1155/2014/180549.

About the author

NITAR NWE, biochemist was born in Mawlamyine, Myanmar on 1971. She is daughter of Mr. Hla Pe and Mrs. Hla Thein. She received BSc in Chemistry with honours from Yangon University, Myanmar on 1995. She received MSc from Bioprocess Technology Program, School of Environment, Resources and Development, Asian Institute of Technology, Pathumthani, Thailand on 1997. She received PhD in Technical Science from Bioprocess Technology Program, School of Environment, Resources and Development, Asian Institute of Technology, Pathumthani, Thailand on 2002. From 1997 to 2004, she was as a Researcher in Bioprocess Technology Program, School of Environment, Resources and Development, Asian Institute of Technology, Pathumthani, Thailand. From 2004 to 2006, she was as a Research Officer in Temasek Life Science Laboratory, National University of Singapore, Singapore. From 2006 to 2010, she was as a Postdoctoral fellow in Faculty of Chemistry, Materials and Bioengineering, Kansai University, Osaka, Suita, Japan. From 2007 to 2009, she was a JSPS Postdoctoral Fellow, Japan Society for the Promotion of Science, Tokyo, Japan. From 2011–present, she is a Founder and Research Scientist of Dukkha Life Science Laboratory (under registration process), Thanlyin, Yangon, Myanmar. From 2012 to present, she is a Consultant in Shan Maw Myae Company, Yangon, Myanmar. From 2013 to 2014, she was a trainer and at present she is a Research Scientist in an Ecological Laboratory, Advancing Life and Regenerating Motherland, Yangon, Myanmar. Her research interest is production of biopolymers from natural resources and applications of biopolymers.

www.ingramcontent.com/pod-product-compliance
Lightning Source LLC
Chambersburg PA
CBHW050452190326
41458CB00005B/1247
9780791860403